Corrosion Science: Modern Trends and Applications

Edited by

N. Suresh Kumar

Department of Physics
JNTU College of Engineering
Anantapuramu-515002
Andhra Pradesh
India

P. Banerjee

Department of Physics
GITAM Deemed To Be University
Bengaluru Campus
Bengaluru
India

H. Manjunatha

Department of Chemistry
GITAM Deemed To Be University Bengaluru Campus
Bengaluru -562163, Karnataka
India

&

K. Chandra Babu Naidu

Department of Physics
GITAM Deemed To Be University
Bengaluru Campus
Bengaluru
India

Corrosion Science: Modern Trends and Applications

Editors: N. Suresh Kumar, P. Banerjee, H. Manjunatha & K. Chandra Babu Naidu

ISBN (Online): 978-981-14-8183-3

ISBN (Print): 978-981-14-8181-9

ISBN (Paperback): 978-981-14-8182-6

need for a court order if at any point you breach any terms of this License Agreement. In no event will any delay or failure by Bentham Science Publishers in enforcing your compliance with this License Agreement constitute a waiver of any of its rights.

3. You acknowledge that you have read this License Agreement, and agree to be bound by its terms and conditions. To the extent that any other terms and conditions presented on any website of Bentham Science Publishers conflict with, or are inconsistent with, the terms and conditions set out in this License Agreement, you acknowledge that the terms and conditions set out in this License Agreement shall prevail.

Bentham Science Publishers Pte. Ltd.
80 Robinson Road #02-00
Singapore 068898
Singapore
Email: subscriptions@benthamscience.net

BENTHAM SCIENCE

CONTENTS

PREFACE

According to the European Federation of Corrosion, the term corrosion signifies an irreversible process due to an interaction with the environment, which causes detrimental effects on the materials permanently. For any industrial applications, this process has a severe impact on revenue. For our health and environment, it can damage several vital sections. There is a severe impact on the GDP of the country as well due to the overall damages. In this sense, the investment of scientific talent for the control of corrosion is always economically justified for mankind.

The overall growth of the Industry 4.0 and subsequent demand for new innovative materials opens a new field of mechanism to control premature degradation of the material. This book entitled 'Corrosion Science-Modern Trends and Applications' with twelve high-quality chapters provided the required picture of the recent contribution and development in the field of corrosion.

Chapter 1 presents the accelerating corrosion test method with various operating apparatus designed by ASTM standard guide. Another most important use of electrochemical techniques, namely linear polarization, potentiodynamic polarization and electrochemical impedance spectroscopy, discusses measuring the corrosion rate for coated and painted specimens.

Chapter 2 discusses various anticorrosive coatings, namely barrier, inhibitive, sacrificial, inorganic, and organic coatings. The various modified coatings are introduced to reduce pores and defects observed in conventional coatings.

Chapter 3 reviews corrosion of electronic printing board with the possibilities and types of the corrosion as well as their protective phenomena. The chapter highlights the corrosion in electronic gadgets and printed electronic circuit board since nowadays, which have printed electronic circuit board.

Chapter 4 contains a discussion on protective materials like conductive polymer and its composites such as conducting polymers mixers of magnetic materials and graphene additives on corrosion resistance capabilities.

Chapter 5 discusses corrosion protection in drinking water systems. The chapter presents corrosion of drinking water distribution system (DWDS) based on various parameters like pH of water, the hardness of water, alkalinity of water, buffer intensity, total dissolved oxygen and total inorganic carbon and organic carbon.

Chapter 6 concentrates on the corrosion of concrete. The electrons can move in the steel rebar, and the ions can move in the concrete, which acts as an electrode leading to the corrosion in concrete. To know the damage of the reinforcement in the concrete X-ray microcomputed tomography is employed.

Chapter 7 presents the corrosion effect on the aluminium alloy. The damage caused by the Sulfate Reducing Bacteria (SRB) is discussed here. The microstructure cracking occurs due to many factors like corrosion, mechanical stress, thermal stress and bacterial adherence.

Chapter 8 focuses on the corrosion problems of nuclear waste systems because of the disposal. The discussion between the rate of corrosion of all the nuclear waste packages with the nuclear waste disposal concept and the safety measures of the landfill sites is featured. Furthermore, the corrosion of the various kinds of nuclear waste packages, and the metallic container for the high-level waste packages review given the deterioration or dissipation processes, the experimental in-situ approaches, and the exemplary of corrosion of nuclear waste and lifetime forecasting are presented in this chapter.

Chapter 9 presents the impact of Microbiologically influenced corrosion (MIC) in various fields like industries related to healthcare, marine, petroleum, and oil. An attempt is made to present MIC and its underlying mechanisms. Cathodic depolarization theory, along with the other mechanisms supporting MIC caused by sulfate/nitrate-reducing bacteria, has been the focus of this chapter.

Chapter 10 recognizes corrosion as a severe problem in the power plant sector. Corrosion gives rise to wastage of material in huge quantities, failure of tubes, leakage of tubes, sudden shutdowns as well as the reduction in the lifetime of components. Also, it reduces the thermal and electrical efficiency of a power plant to a maximum extent leading to minimum maintenance, outage, and replacement cost.

Chapter 11 depicts the impacts of corrosion on significant industries in chemical and fertilizers industries. In this chapter, they focus on corrosion related to chemical and fertilizer industries, the impact of corrosion on their efficiency, corrosion controlling methods and their interrelated phenomena.

Chapter 12 focuses on Marine corrosion, its mechanism, factors affecting corrosion, and several methods adopted for the prevention of corrosion describes with an emphasis on marine corrosion inhibitors. The use of inorganic compounds and paints as corrosion inhibitors is discussed in the chapter.

K. Chandra Babu Naidu
Department of Physics
GITAM Deemed To Be University, Bengaluru Campus
Bengaluru-562163, Karnataka
India

List of Contributors

A. Manohar   Department of Materials Science and Engineering, Korea University, 145 Anam-ro Seongbukgu, Seoul 02841, Republic of Korea,

A. Franco   Instituto de Física, Universidade Federal de Goiás, Goiânia, Brazil

B. Parvatheeswara Rao Department Of Physics, Andhra University, Visakhapatnam-530003, India

B. Venkata Shiva Reddy Department of Physics, The National College, Bagepalli-561207, Karnataka, India

D. Chandra Sekhar Department of Physics, Govt. College (A), Anantapuramu-515002, A.P, India

D. Baba Basha   Department of Physics, College of Computer and Information Sciences, Majmaah University, Al'Majmaah, Saudi Arabia

H. Manjunatha   Department of Chemistry, GITAM Deemed to be University, Bangalore - 562163, Karnataka, India

K. Ram Mohan Rao Department of Chemistry, GITAM Deemed to be University, Visakhapatnam 530045, Andhra Pradesh, India

K. Haripriya   Department of Anaesthesia, MIMS Medical College, NTR University, Vijaywada, Andhra Pradesh, India

K. Chandra Babu Naidu Department of Physics, GITAM Deemed to be University, Bangalore - 562163, Karnataka, India

K. Venkata Ratnam Department of Chemistry, GITAM Deemed to be University, Bangalore - 562163, Karnataka, India

K. Chandra Babu Naidu Department of Physics, GITAM Deemed to be University, Bangalore - 562163, Karnataka, India

M.S.S.R.K.N. Sarma Department of Physics, GITAM Deemed to be University, Bangalore - 562163, Karnataka, India

M. Ajay Kumar   Seongbuk-gu, Seoul 02841, Republic of Korea, Seongbuk-gu, Seoul 02841, Republic of Korea

M. Balaraju   Department of Chemistry, PSC & KVSC Govt. Degree College, Kurnool-518502, A.P, India

N. Suresh Kumar   Department of Physics, JNTU College of Engineering Anantapur, Anantapuramu-515002, A.P, India

N.V. Krishna Prasad Department of Physics, GITAM Deemed to be University, Bangalore - 562163, Karnataka, India

N. Suresh Kumar   Department of Physics, GITAM Deemed to be University, Bangalore - 562163, Karnataka, India

U. Naresh   Department of Physics, BIT Institute of Technology, Hindupur, A.P, India

T. Anil Babu   Department of Physics, GITAM Deemed to be University, Bangalore-562163, Karnataka, India

P. Banerjee   Department of Physics, GITAM Deemed to be University, Bangalore campus, Visakhapatnam 530045, Andhra Pradesh, India

S. Ramesh   Department of Physics, GITAM Deemed to be University, Bangalore - 562163, Karnataka, India

CHAPTER 1

Testing the Types of Corrosion

D. Chandra Sekhar[1], N. Suresh Kumar[2,*], K. Chandra Babu Naidu[3], B. Venkata Shiva Reddy[3,4] and T. Anil Babu[3]

[1] *Department of Physics, Govt. College (A), Anantapuramu-515002, A.P., India*

[2] *Department of Physics, JNTU College of Engineering Anantapur, Anantapuramu-515002, A.P., India*

[3] *Department of Physics, GITAM Deemed to be University, Bangalore - 562163, Karnataka, India*

[4] *Department of Physics, The National College, Bagepalli-561207, Karnataka, India*

Abstract: This chapter presents the accelerating corrosion test method with various operating apparatus designed by ASTM standard guide. It is used to measure the corrosion resistance of coated and painted samples. The important uses of electrochemical techniques are linear polarization, potentiodynamic polarization, and electrochemical impedance spectroscopy. These are discussed in measuring the corrosion rate for coated and painted specimens.

Keywords: Corrosion, Electrochemical Cell, Electrochemical Impedance Spectroscopy, Salt Fog Test.

1. INTRODUCTION

In the early 1900s, the accelerated corrosion test method of metallic coatings was initially established for testing the samples, thereby improving the efficiency and durability. The test processes for corrodibility of nonferrous, ferrous metals, inorganic, and organic coatings were increased. The changes observed were added for useful testing of new materials. The service durability, and quality of samples were tested using accelerated corrosion tests. Further, these were used in material research. The ASTM International, the Society of Automotive Engineers (SAE), the Federation of Societies for Coatings Technologies (FSCT), and others established many accelerated corrosion tests, and provided variations in technical fields as well as industrial materials. As the water-based coating technology continues to evolve, one of the significant challenges that increases their anti-

* **Corresponding author N. Suresh Kumar:** Department of Physics, JNTU College of Engineering Anantapur, Anantapuramu-515002, A.P., India; Tel: +91-81211 27157; E-mail: sureshmsc6@gmail.com

N. Suresh Kumar, P. Banerjee, H. Manjunatha and K. Chandra Babu Naidu (Eds.)

corrosion performance. Some powder coatings, and their application methods were developed to decrease the usage of solvents, and applications. Subsequently, the material market made a drastic increase in the coatings, and the surface coatings were proven to have longer warranties for market challenges. The oldest accelerated corrosion testing method was ASTM B 117 standard operating Salt Spray (Fog) Apparatus. A recent development was used to measure the corrosion rate of metallic coatings in a "near-shore" atmosphere. A corrosion test was applied to determine the efficiency of material samples, and furthermore, these were used in the development, and modifications for applications in the environmental, and material sciences. Hence, the corrosion rate of the test specimen was measured using the several types of accelerated corrosion testing methods to be described at present.

2. SALT-SPRAY (FOG) TEST

The corrosion resistivity can be determined by using the Standard test ASTM B 117 for inorganic, and organic coatings applied to the metals [1]. The B 117 Standard was made up of static (constant) condition, and continuous test, which was ideally carried out in multiples within a 24-hour period. The cabinet consists of an atomized solution, which was prepared by using 5% sodium chloride, and 95% by mass of ASTM D 1193 Type IV water. This kind of cabinet can be called a salt Fog chamber. The test samples were exposed to this salt fog environment. The chamber contains an atomized solution which was held at a temperature of 35°C, and 95% relative humidity. To preserve these conditions, the chamber was heated, and it was also necessary to maintain wet conditions at the bottom of the exposure zone. The corrosion resistance of various Al and Al/Zn-coated AZ91D Mg alloys was estimated by the salt spray method. After exposing it within the salt spray chamber for one hour, fast corrosion can be occurred on the bare surface of Mg alloy, causing the loss of metallic lustre. After 8 hr of salt spray test, the Al coated Mg sample maintains its integrity, and the appearance of silver-grey surface indicating improved corrosion resistance [2].

3. MODIFIED SALT FOG TESTS

3.1. Acetic Acid Salt Spray (Fog) Test

ASTM G 85 was observed to be the modified salt fog test, which includes five annexes. The primary one is the acetic acid-salt fog test, in which the salt solution was prepared according to the B 117. Afterwards, the pH values were maintained within the range from 3.1 to 3.3 to the acetic acid. The acetic acid-salt spray test can be used for decorative chromium plating on steel [3]. The test duration should be 144-240 h. It can be used to test the metallic coatings, inorganic, and organic coatings for resistance to the highly corrosive environment in comparison with the ASTM B 117.

3.2. Cyclic Acidified Salt Fog Test

In the case of the second annex of the cyclic acidified salt fog testing, the ASTM G 43 was used for exfoliated tests on certain aluminium alloys [3]. It was generally indicated as the MASTMAASIS test, which can be used for exfoliation testing of aluminium. This test uses 5% of sodium chloride solution, and it makes it easy to adjust pH in the range of 2.8 to 3.0 with acetic acid. The test duration was noted to be 6 h cycle of ¾ h spray; 2 h dry air purge and ¾ h soak at high humidity. The temperature of chamber exposure was held at 49°C, and the tower of humidity was operated at 57°C. During the purge cycle, the objective of the test can be the drying process of samples, and further leaving a white corrosion product.

3.3. Acidified Synthetic Sea Water (Fog) Test

In annexe 3, this testing can increase the application for manufacturing the control of exfoliation-resistant heat treatments used in producing 2, 5, and 7 K series of aluminium alloys. Then, the pH value was observed to be 2.5 - 3.0, and the testing method was operated at 49°C. The modified method was used to test the corrosion rate of organic coatings applied on metallic substrates while the testing procedure was operated between 24 to 35°C. At the time of this test operated at 1 to 2 ml/h, the collection rate specified for fog cycles was the same as the use of B 117 Standard. However, 2 h cycles were used for the whole test period. Due to the cyclic nature of this test, it was necessary to develop and confirm proper condensate collection rates by using 16 h salt fog test to be conducted periodically [4]. The apparatus in the test chamber must be controlled, and hence that will cycle the exposure zone run throughout ½ h salt spray, 1 ½ h of soaking time at 98% of relative humidity.

3.4. Salt/SO$_2$ Spray (Fog) Test

Standard ASTM G85 Annex 4 is a salt/SO$_2$ fog (spray) test, which can be performed using either sodium chloride solution or synthetic sea salt solution. Like the salt spray test, it was performed at 35°C. The fog may be continuous or infrequent. In case of the operation of the cycle during the test period, the sulfur dioxide can be injected into a chamber. Herein, for ½ h salt fog, and ½ h SO$_2$ were injected into a chamber. These were soaked for 2 h. Hence, the Navy developed the same kind of test used to find the exfoliating corrosion on aircraft carriers [5].

3.5. Dilute Electrolyte Cyclic Fog/Dry Test

0.35% of ammonium sulfate, and 0.05% of sodium chloride in 0.60% by mass of ASTM D 1193 Type IV water were used to prepare the electrolyte solution, which was necessary to perform the test. When compared with the standard salt fog test, this solution was found to be more dilute and was operated using 2 h cycle time. This consists of 1 h fog at ambient 24°C ± 3°C and less than 75% of relative humidity followed by 1 h dry off at 35°C. Hence, at controlled room temperature, the test samples were exposed to 1 h of salt fog. Afterwards, the sample was dried at 35°C for one hour. The pH value of collected condensate falls in the range of 5.0 to 5.4. Due to the cyclic nature of this test, a distinct salt fog test with a duration of 16 h can be required to develop and confirm good collection rates. Due to the changes in humidity of this testing procedure, and the cyclic salt test, the cabinet needs a valve that allows the atomized air to bypass the humidifying tower, and time devices to control the duration of cycle, spray, temperature variations, and airflow. This was noticed as a modified British Rail "Prohesion" testing, which was established in the 1960s for the coated metals in industrial sectors [6]. This kind of test method was used to test the paints and coatings on steel materials.

4. CYCLIC SALT FOG/UV EXPOSURE

The painted metals were exposed to salt for/UV environment and tested using standard ASTM D 5894. This was incorporated with a cycle of UV exposure followed by G 85 Annex 5 exposure zone [7]. The cycle period of UV exposure introduced environmental impact to the condition of test, and originated results which were like the direction of the field. Two distinct cabinets are required instead of the salt fog/UV chamber.

5. CASS TEST

The ASTM B 368 Standard [8] was established by the American Electroplaters, and Surface Finishers Society (AESF) [9], which was used initially in the improvement of metallic coatings, in addition to decorative coatings. These were exposed to highly corrosive environments. It was observed to be a significant challenge to provide a prompt evaluated service for testing the specification of the product in research studies. It became necessary to provide controlled manufacture possessing the environment friendly in nature. This type of modification can be done with human effort. The modified primer B117 salt spray test is the standard B 368. The transformation can be done by adjusting the pH of the 5% salt solution within the range of 6-7, and then adding 0.25 g of copper chloride reagent per liter of salt solution. This test can be operated at the temperature in the exposure zone of 49°C. This runs continuously for 6 to 170 hours as admitted between seller and purchaser. During regular days, it was noted to be necessary to check the temperature within the exposure zone. That is, for "short" duration, the chamber must be opened periodically to operate the samples and reload solution. Another modified salt fog test can become the consistency of test exposure circumstances. Mass loss of nickel coupons was used in the standard ASTM B 368 rather than the steel used in the B 117 standard. The corrosion test chamber devices equipped with the B 368 standard achieve the need for the Standard ASTM B 117 and will resist the increasing temperature. This test uses a powerful electrolyte solution.

5.1. Corrodkote Test

ASTM B 380 [10], corrodkote test was established by AES to enhance the efficiency of decorative painting used in the industry of automobile. The corrodkote test uses a mixture contains ferric chloride, kaolin, cupric nitrate, and ammonium chloride (NH_4Cl) in the distilled water [11]. The test uses those samples which were practiced with the mixture. These were injected in the humidity chamber maintained at 38°C, and 80 and 90% of relative humidity without condensation on parts. The test period was about 24 h per cycle [11]. A final process of the test, and the cleaning of samples were done by using moderated working water. To get the final corrosion product, these cleaned samples were needed to exposure to salt fog or humidity.

5.2. Filiform Test

The corrosion resistance of metal with organic coatings can be tested by using the Filiform test as standard ASTM D 2803 [12]. According to ASTM B 117, the test begins with a salt exposure immediate after 70 to 90% of relative humidity which needs to be maintained for the exposure of humidity. The differences in the test can produce the change in the duration of the salt fog exposure, and the amount of relative humidity percentage. The test method of filiform corrosion for the painting on the aluminum wheels, and aluminum wheel trim was incorporated with the CASS test. This test exposure for 6 h before the humidity cycle [13]. Filiform type corrosion was recognized by thread-like filaments as obtained under clear as well as a coated aluminum powder [14].

5.3. High Humidity Tests

ASTM D 2247, Standard practice for testing water resistance of coatings in 100% relative humidity was widely used. The water-resistance of coated samples was tested using this method. The temperatures of both cabinet, and humidity tower were fixed at 38°C. The device should be a water container with an electrical heating element or test container with pipe assembly arrangement for dispersion of air. Due to this kind of built, it makes condensation occurring on the samples. In case of the modified salt fog chamber B 117, first, remove the fogging apparatus, and then replaced by a water container or the pipe assembly arrangement for dispersion of air. The water height in the water container with the heating process can be decreased to a position above within the humidifying water.

5.4. Corrosive Gas Tests

The corrosive gas test can be treated as a mixture of gas placed into a large humidity environment, and widely used for porosity test. The following examples are listed of these tests:

- The moisture SO_2 test can practice using standard ASTM G 87.
- Blended Flowing gas test can be used by standard ASTM B 827.
- In the presence of SO_2, the saturated humid atmospheres have used the test such as standard DIN 50018.

6. ASTM G 87

The moisture SO_2 test operated by ASTM G 87, and later, it was modelled by the DIN 50018, and is often called the test of Kesternich. The low concentrations of SO_2 gas were added into 100% of the humid atmosphere for conducting the test. A temperature of 40°C was required to maintain this test. The DIN 50018, and G 87 were used in electronic industrial society, as well as on both secured devices (ASTM D 6294) [15], and roofing materials. A recent study for oil coatings (D 3794) test was also conducted by G87. Due to the other gases such as carbon dioxide lead the variation in the trial.

6.1. Mixed Flowing Gas

ASTM B 827, Standard practice for conducting the blended flowing gas test can be widely used in the electronic industry. The testing method uses a combination of various gases at controlled temperature, and humidity to produce the required conditions. The MFG test requires equipment of a more complex nature than corrosive gas tests.

7. CYCLIC CORROSION TESTS

The corrosion tests include a single step or more steps involved for salt fog exposure, humidity, and drying, ambient temperature. SAE J2334, Standard practice for cosmetic corrosion lab testing, was used in the automotive industry to produce good results in response to field exposure. The test method is:

- The humid stage at 50°C and 100% of humidity meet at 6-h.
- At ambient conditions, salt fog exposure duration is 15 min.
- The dry period at 60°C and then 50% of humidity maintained for 17 h and 45 min, respectively.

The test duration is 60 cycles, and longer cycles were used for heavier coating systems. The use of corrosion coupons was incorporated in the test method for monitoring the corrosion activity within the chamber. A minimum of six coupons made of 25.4 x 50.8 mm AISI 1006-1010 steel was used. The coupons are cleaned, weighted, and mounted in a nonmetallic rack. One coupon from each end of the rack was removed, cleaned, and reweighted every 20 cycles. At present, zero mass loss numbers for each cycle were discussed. The evaluation of the corrosion resistance of metals in the presence of environments will be more effective for chloride ions. For instance, sodium chloride from a seawater source. Due to the cyclic nature of this test, the test samples were exposed to modification of atmospheres during the time duration.

8. ELECTROCHEMICAL TECHNIQUES FOR CORROSION TESTING

Most the industrial markets using the technique of electrochemical corrosion can be suitable for the evaluation of service duration of metallic coatings. This technique was used to determine the corrosion resistance and improving the factors to protect from corrosion. In addition, the environment was characterized by oxidizing power. The environment contains a strong tendency of oxidizing materials. The direct current dependence of electrochemical processes was demonstrated in this chapter. Since linear polarization, and Potentiodynamic polarization techniques used the application of direct current. Electrochemical impedance spectroscopy (EIS) is another technique using the application of alternating current. This determines the frequency dependency mechanism in the process of corrosion and measures the variation in polarization resistance.

In case of the linear polarization method, within the potential range of 10 - 20 mV, the current was determined from the anodic to the cathodic direction. At the potential corrosion range, the current against potential plot shows linearity nature. The calculated slope of this plot gives the resistance of polarization. The pre-determined of anodic and cathodic Tafel constants in the Stern-Geary equation applied to calculate the corrosion current. Potentiodynamic polarization technique was applied to study the corrosion properties of passivating metals, and alloys. Electrochemical Impedance Spectroscopy determines the reaction of corroded systems to the "ac" perturbation potentials. Nowadays, most and wide use of this technique measures the corrosion resistance, and passivation in metals. The study of inhibitors was to prevent corrosion, properties of a sacrificial, barrier, and efficiency of polymer-based coatings like paints.

8.1. Linear Polarization Method-Evaluation of Corrosion Rates

Linear polarization resistance technique measures the rapid corrosion rates, and it uses the environment like the ionic conducting liquid. The same kinds of materials were required to prepare two or three cylindrical electrodes in the LPR probe. The smallest d.c. potential ΔE of 20 mV is applied between two electrodes. The flow of current Δi across the two polarized electrodes was measured after a short duration. The linear polarization resistance R_p was measured from the ratio of voltage to current, and from the Stern-Geary equation, the corrosion current shows inversely proportional relationship to R_p.

$$R_p = (\Delta E / \Delta i) = B / I_{cor} \tag{1}$$

The corrosion rate is calculated from I_{cor} using Faraday's law if the constant B is known.

B can be electrochemically measured by using the anodic and cathodic Tafel slopes.

In a three-electrode probe, the test sample was considered as the first electrode, the second one was an additional electrode, and the last and third one could be reference electrodes opposite to the potential of the sample. The current will be passed through the test electrode, and the additional electrode. The ohmic drop of electrode probe could be responsible for reducing the errors. In addition, it was applicable for low ionic conductivity liquids. Numerous linear polarization resistance techniques with a signal having a high frequency can be used to determine the ohmic (IR) drop across two or three-electrode probes. The use of the LPR probes with suitable conductive to only conductive media like pure water.

Many studies on linear polarization were reported to observe the corrosion resistance, inhibitors that prevent corrosion, the study of passivation of metals; properties of sacrificial type barrier, and efficiency of polymer-based coatings due to the deposition on the metallic samples [16 - 22]. Mennenoh and Engel reported [16] that the use of polarization technique capable to examine the strong inhibitors can prevent corrosion on steel in baths. Linear polarization techniques were used [17 - 19] to measure the resistance of polarization, and the rate of corrosion for several barriers and sacrificial type coatings. Electrodeposited zinc-nicke--cadmium coatings usually contain high polarization resistance (low rate of corrosion) as compared with the other coatings examined in 0.5 M H_3BO_3+0.2M Na_2SO_4 solution at pH of 6.5 [18]. Electrodeposited Zn-Co coatings offer high corrosion resistance (higher R_p value) as compared with pure Zn coatings.

8.2. Potentiodynamic Polarization Measurements

When alloy type or metallic sample is exposed to a specific atmosphere, they become passive at a specified potential region and that potential can be measured using the potentiodynamic polarization technique. It estimates the capability of the metal to passivate immediately and passivation of metals causes the critical current density. The metal and alloys became passive when they were exposed to a corrosive atmosphere. The corrosion rate of this passive metal and alloys can be measured by using potentiodynamic polarization technique. This technique allows the estimation of the active corrosion region, the current density, the commencement of passivation, the initial potential for passivation, the flow of current across the passivation region, and the potential period over the passivated region. Nanostructured ZrO_2-TiO_2 multi-layer composite coating was deposited on AISI 304 stainless steel and its corrosion rate in simulated marine water was evaluated by using potentiodynamic polarization test. The corrosion resistance was increased with increasing thickness coating. Moreover, the increase in corrosion in potential can be responsible for low corrosion current.

8.3. Electrochemical Impedance Spectroscopy (EIS)

The electrochemical cell can be made from three electrodes. First one is the test sample which can act as a working electrode. The coated/uncoated test sample must have surface exposure with an area of cross section of 1-cm^2. The second one is the reference electrode which is the saturated calomel electrode (SCE), and the last one is the cylindrical shape platinum electrode which may serve as a counter electrode. Further, 3.5 weight % of NaCl aqueous solution was added into the electrochemical samples, and the measurements were performed at the temperature 20°C. Measurements by the EIS were operated at the region of the corrosive potential of the test samples within the frequency range from 0.01 to 10,000 Hz. The EIS measurement device uses the sinusoidal voltage (amplitude: ±10 mV). Hence, the EIS measurement data were analyzed in terms of Bode (logarithm of the impedance modulus |Z| against phase angle as a function of the logarithm of the frequency f) diagram and Nyquist plots. The impedance spectra can be analyzed through the model of equivalent circuits.

8.4. Application of Electrochemical Impedance to Corrosion Studies

The electroless nickel-zinc-phosphorous and electrodeposited nickel-zinc alloy provide a magnitude 2 KΩ resistance of barrier. The resistance of barrier coatings was improved by increasing the nickel concentration during the deposition. Since the strength of corrosion protection was decided using organic coatings and the service duration of these coatings measured with the application of the EIS technique [23 - 38]. It can be a potential technique to collect special parameters of the system to determine the primary period of their response and degradation in the properties of barrier coatings [23, 36, 37]. The variation in the coating capacitance was identified and correlated with the water [31]. The variation in the resistance can be expressed as the permeability of ionic specimens in the electrolyte [39 - 41]. The EIS can be capable of measuring the separation into constituent layers of the epoxy layer electro-coated on the phosphate applied for steel in 3.5% sodium chloride solution which was exposed into the atmosphere [29]. The parameters of coatings were calculated using an analogue circuit model. The area of separated layers was determined by the resistance of pore and the frequency of break point. The measurements on an area of corrosion comparable with obtained. The properties of barrier epoxy/amine and epoxy/ phenolic coated soft steel can be estimated with EIS when a specimen is exposed to 3.5% sodium chloride solution in the atmosphere [30].

8.5. Advantages and Limitations of EIS

In a single step cycle of performance, EIS became a strong technique to study the properties of electrochemistry and chemistry at the interface/barrier of metal-electrolyte [42, 43]. Even if a small ac-current or voltage is applied in the EIS technique, it does not have a significant effect on electrochemical activity in an electrochemical cell. The information is well known if the process occurs in the electrolytes with low conductivity. Since this technique uses direct current processes without free error, the potential can estimate the phenomenon at the interface between metal and electrolyte. Hence, this technique is used to measure the capacitance of electrode, the mechanical processes of electrochemical activity and corrosion on the kinetics of charge transfer. EIS technique used to measure the resistance of polarization with more accuracy. The use of multi-frequency excitations made possible measuring the capacity of processes of corrosion at distinct rates. Bode and Nyquist plots were used for the analysis of data related to impedance. Moreover, it can be necessary to obtain the same results of ac for Bode diagrams and an analogue circuit. It can be achieved by adjusting the resistances and capacitances.

9. RECENT CORROSION RESEARCH

At grain boundaries, the localized corrosion like cracks and pits can be used for an evaluation of such corrosion rates which is quite difficult using the EIS technique [44]. Recently, the estimation of localized corrosion process was done using scanning electrochemical microscopy (SECM), localized electrochemical impedance spectroscopy (LEIS), and scanning Kelvin probe (SKP) [45, 46]. Subsequently, these techniques were used to evaluate the properties of the passive film [47, 48]. Spatial resolution electrochemical spectroscopy dependency of LEIS technique was used to measure the corrosive resistant materials capacitance to inhibit the occurrence of corrosion process. These were wear testing corrosion, pit growth corrosion, and coating degradation. This method was used to measure the variations during the growth of pit corrosion [49, 50]. The authors reported that the limitations in spatial resolution and the mathematical formulation of this method can be controlled using the known parameters [51 - 56]. Even though numerous results were reported in this field, the illustration of the localized corrosion mechanism can be understood as well in the present cases [57].

The capacitance of coating may lead the primer variation in the properties of the coating, and it can be determined using the EIS technique [58]. The previous scientists reported that [59] the protection polymer coatings were studied using EIS tool. Moreover, the data of EIS can be analyzed, and simulated by various methods. In the case of protective coating dopant, element dependency model, it introduced the impedance of Warburg [60] to plot the spectra of electrochemical impedance. The literature [61] describes the applications of EIS for the estimation of the properties of anticorrosive coatings and the studies on inhibitors to prevent corrosion in metal and alloy-based coatings, passivated films. The corrosion behavior can be studied using the EIS method in various fields like biomedical, reinforced concrete, and microbiological. At certain dynamic conditions, the modern impedance technique was used, which is known as dynamic electrochemical impedance spectroscopy (DEIS) [62, 63]. Since it depends on modern dynamic electrochemical techniques like pulse cyclic voltammetry, this powerful technique can be used to measure the rate of generation and repeated passivation of pits with intermediate stable.

CONCLUSION

This chapter explained the various accelerated corrosion tests, namely salt fog, modified salt spray, acidified test, salt/SO_2 test, CASS test, and cyclic corrosion test. The test specimen was corroded under salt spray exposure. Electrochemical techniques were employed to coated and painted samples to measure the corrosion resistance. Linear polarization method measures the polarization resistance of coated specimen which can be used as a working electrode in an electrochemical cell. OCP *vs.* time measurements of the samples revealed that the pit corrosion potential was decreased while increasing OCP. The potentiodynamic polarization estimates the corrosion resistance of coated test samples using electrochemical cell. The EIS technique can be the response of the electrochemical cell to the applied ac-potential. The impedance is, in general, the ratio of ac-voltage to ac-current. The Bode and Nyquist plots of tested samples were analyzed through equivalent circuits which were used to measure the corrosion resistance.

CONSENT FOR PUBLICATION

Not applicable.

CONFLICT OF INTEREST

The author declares no conflict of interest, financial or otherwise.

ACKNOWLEDGEMENTS

Declared none.

REFERENCES

[1]   ASTM Standard B 117-97. Standard Practice for Operating salt spray (Fog) Apparatus, Annual Book of ASTM Standards, ASTM International *West Conshohocken, PA,* **2001**, 3-2.

[2]   ASTM Standard G 85-98. Standard practice for modified salt spray (fog) testing, Annula book of ASTM standards, ASTM international *West Conshohocken, PA,* **2001**, 3-2.

[3]   *Corrosion tests and standards: Application and interpretation, MNL 20, R.Baboian, Ed.ASTM international West Conshohocken, PA,* **1995**.

[4]   Sprowls, D.O. Exfoliation chap. 22 in corrosion tests and standards: Application and interpretation, MNL 20. ASTM international West Conshohocken*ASTM international West Conshohocken*; Baboian, R., Ed.; , **1995**, p. 220.

[5] Cremer, N.D. Prohesion compared to salt sprayand outdoors *presented at Federation of societies for coatings technology Paint industries show,* **1989**.

[6] *ASTM standard D 5894: Standard practice for cyclic salt fog/UV exposure of painted metal, annual book of ASTM standards*; ASTM International: West Conshohocken, PA, **2001**, pp. 6-01.

[7] Pinner, W. Accelerated Corrosion tests for the performance of plated coatings *AESF research project No. 15, Fourth progress report,*

[8] ASTM Standard B 368. Standard method for copper-accelerated acetic acid-salt spray (fog) testing (CASS test), Annual book of ASTM Standards *ASTM International, West Conshohocken, PA,* **2001**, 3-2.

[9] ASTM Standard B 380. Standard method of corrosion testing of decorative electrodeposited coatings by the corrodkote procedure, Annual book of ASTM Standards *ASTM International, West Conshohocken, PA,* **2001**, 3-2.

[10] SAE J2334. Lab cosmetic corrosion test *Society of Automotive Engineering surface vehicle standard,* **1998**.

[11] ASTM Standard D 2803. Standard guide for testing Filform corrosion resistance of organics coatings on Metal, Annual book of ASTM Standards *ASTM International, West Conshohocken, PA,* **2001**, 6-1.

[12] GM9682 General Motors Engineering Standard. *Filiform Corrosion test procedure for painted Aluminum wheels and painted aluminum trim,* **1997**.

[13] Coyle, J.; Fortier, M.; McKeon, J. *studying Filiform corrosion of powder coating,* **1997**, 22-35.

[14] ASTM Standard D 6294. Standard test method for corrosion resistance of Ferrous metal fastener Assemblies used in roofing and waterproofing *Annual Book of ASTM Standards, ASTM International, West Conshohocken, PA,* **2001**, 1-8.

[15] ASTM Standard D 3794. Guide for Coil coatings, Annual book of ASTM Standards *ASTM International, West Conshohocken,* **2001**, 6-1.

[16] Mennenoh, S; Engel, H.J. A method for measuring the effectiveness of inhibitors in pickling baths *Stahl Eisen,* **1962**, *82*, 1796-1801.

[17] Veeraraghavan, B.; Kim, H.; Haran, B.; Popov, B. Comparison of mechanical, corrosion and hydrogen permeation properties of electroless Ni-Zn-P alloys with electrodeposited Zn-Ni and Cd. *Corrosion,* **2003**, *59*, 1003-1010.
[http://dx.doi.org/10.5006/1.3277518]

[18] Kim, H.; Popov, B.; Chen, K.S. Comparison of corrosion resistance and hydrogen permeation properties of Zn-Ni, Zn-Ni-Cd and Cd coatings on low carbon Steel. *Corros. Sci.,* **2003**, *45*, 1505-1521.
[http://dx.doi.org/10.1016/S0010-938X(02)00228-7]

[19] Veeraraghavan, B.; Kim, H.; Popov, B.N. Optimization of electroless Ni-Zn-P deposition process: experimental study and mathematical modeling. *Electrochim. Acta,* **2004**, *49*, 3143-3154.
[http://dx.doi.org/10.1016/j.electacta.2004.01.035]

[20] Durairajan, A.; Haran, B.S.; White, R.E.; Popov, B.N. Development of a new electrodeposition process for plating of Zn-Ni-X (X1/4Cd, P) alloys: Corrosion characteristics of Zn-Ni-Cd ternary alloys. *J. Electrochem. Soc.,* **2000**, *147*, 1781-1786.
[http://dx.doi.org/10.1149/1.1393434]

[21] Ganesan, P.; Kumaraguru, S.P.; Popov, B.N. Development of Zn-Ni-Cd coatings by pulse electrodeposition process. *Surf. Coat. Tech.,* **2006**, *201*, 3658-3669.
[http://dx.doi.org/10.1016/j.surfcoat.2006.08.143]

[22] Ganesan, P.; Choi, Y.I.; Kumaraguru, S.P.; Popov, B.N. Development of corrosion-resistant silica coatings on surface modified zinc-coated steel. *J. Appl. Surf. Finish.,* **2007**, *2*, 20-28.

[23]   Dickinson, J.J.; Lofti, S. The anodic dissolution of tin in sodium hydroxide solutions. *Electrochim. Acta*, **1978**, *23*, 513-519.
[http://dx.doi.org/10 1016/0013-4686(78)85029-4]

[24]   Mansfeld, F. Don't be afraid of electrochemical techniques—but use them with care. *Corrosion*, **1988**, *44*, 856-868.
[http://dx.doi.org/10.5006/1.3584957]

[25]   Murray, J.N.; Moran, P.G. Influence of moisture on corrosion of pipeline steel in soils using *in situ* impedance spectroscopy. *Corrosion*, **1989**, *45*, 34-43.
[http://dx.doi.org/10.5006/1.3577885]

[26]   Khangholi, A.; Revilla, R.I.; Lutz, A.; Loulidi, S.; Rogge, E.; Van Assche, G.; De Graeve, I. Electrochemical characterization of plasma coatings on printed circuit boards. *Prog. Org. Coat.*, **2019**, *137*105256
[http://dx.doi.org/10.1016/j.porgcoat.2019.105256]

[27]   Lorenz, W.J.; Mansfeld, F. Determination of corrosion rates by electrochemical DC and AC methods. *Corros. Sci.*, **1981**, *21*, 647-672.
[http://dx.doi.org/10.1016/0010-938X(81)90015-9]

[28]   Macdonald, D.D. *Transient Techniques in Electrochemistry*; Plenum Press: New York, **1977**.
[http://dx.doi.org/10.1007/978-1-4613-4145-1]

[29]   Epelboin, I.; Jousselin, M.; Wiart, M. Impedance measurements for nickel deposition in sulfate and chloride electrolytes. *J. Electroanal. Chem.*, **1981**, *119*, 61-71.
[http://dx.doi.org/10.1016/S0022-0728(81)80124-6]

[30]   McCluney, S.A.; Popova, S.N.; Popov, B.N.; White, R. Comparing EIS methods for estimating the degree of delamination of organic coatings on steel. *J. Electrochem. Soc.*, **1992**, *139*, 1556-1560.
[http://dx.doi.org/10.1149/1.2069454]

[31]   Popov, B.N.; Alwohaibi, M.M.A.; White, R.E. Using electrochemical impedance spectroscopy as a tool for organic coating solute saturation monitoring. *J. Electrochem. Soc.*, **1993**, *140*, 947-951.
[http://dx.doi.org/10.1149/1.2056233]

[32]   Scully, J.R. Electrochemical impedance of organic-coated steel: correlation of impedance parameters with long-term coating deterioration. *J. Electrochem. Soc.*, **1989**, *136*, 979-990.
[http://dx.doi.org/10.1149/1.2096897]

[33]   Macdonald, D.D.; Mc Kubre, M.C.H. Electrochemical impedance techniques in corrosion science, **1981**.
[http://dx.doi.org/10.1520/STP28030S]

[34]   Bard, A.J.; Faulkner, L.R. *Electrochemical Methods*; Wiley: New York, **1980**.

[35]   Mansfeld, F.; Kerdig, M.W.; Tsai, S. Application of acoustic emission to detection of reinforcing steel corrosion in concrete. *Corrosion*, **1982**, *38*, 9-14.
[http://dx.doi.org/10.5006/1.3577322]

[36]   Padget, J.C.; Moreland, J.P. Use of A.C. impedance in the study of the anticorrosive properties of chlorine-containing vinyl acrylic latex copolymers. *J. Coatings Technology*, **1983**, *55*, 39-51.

[37]   Kending, M.W.; Mansfeld, F.; Tsai, S. Determination of the long term corrosion behavior of coated steel with AC impedance measurements. *Corros. Sci.*, **1983**, *23*, 317-329.
[http://dx.doi.org/10.1016/0010-938X(83)90064-1]

[38]   Lorenz, W.J.; Mansfeld, F. Corrosion inhibition in neutral, aerated media. *J. Electrochem. Soc.*, **1985**, *132*, 290-296.
[http://dx.doi.org/10.1149/1.2113820]

[39]   Kending, M.W.; Scully, J.R. Basic aspects of electrochemical impedance application for the life prediction of organic coatings on metals. *Corrosion*, **1990**, *46*, 22-29.

[http://dx.doi.org/10.5006/1.3585061]

[40]   Tait, S.W. Using electrochemical measurements to estimate coating and polymer film durability. *J. Coatings Technology,* **2003**, *75*, 45-50.
[http://dx.doi.org/10.1007/BF02730070]

[41]   Haran, B.S.; Popov, B.N.; White, R.E. Determination of hydrogen diffusion coefficient in metal hydrides by impedance spectroscopy. *J. Power Sources,* **1998**, *75*, 56-63.
[http://dx.doi.org/10.1016/S0378-7753(98)00092-5]

[42]   Durairajan, A.; Haran, B.S.; White, R.E.; Popov, B.N. Pulverization and corrosion studies of bare and cobalt-encapsulated metal hydride electrodes. *J. Power Sources,* **2000**, *87*, 84-91.
[http://dx.doi.org/10.1016/S0378-7753(99)00399-7]

[43]   Macdonald, D.D.; Smedley, S.I. Characterization of vacancy transport in passive films using low frequency electrochemical impedance spectroscopy. *Corros. Sci.,* **1990**, *31*, 667-672.
[http://dx.doi.org/10.1016/0010-938X(90)90178-8]

[44]   Macdonald, D.D. Reflections on the history of electrochemical impedance spectroscopy. *Electrochim. Acta,* **2006**, *51*, 1376-1388.
[http://dx.doi.org/10.1016/j.electacta.2005.02.107]

[45]   Macdonald, D.D. Some advantages and pitfalls of electrochemical impedance spectroscopy. *Corrosion,* **1990**, *46*, 229-242.
[http://dx.doi.org/10.5006/1.3585096]

[46]   Katemann, B.B.; Schulte, A.; Calvo, E.J.; Koudelka-Hep, M; Schuhmann, W. Localized electrochemical impedance spectroscopy with high lateral resolution by means of alternating current scanning electrochemical microscopy *Electrochem. Commun,* **2002**, *4*, 134-138.

[47]   Eckhard, K.; Shin, H.; Mizaikoff, B.; Schuhmann, W.; Kranz, C. Alternating current (AC) impedance imaging with combined atomic force scanning electrochemical microscopy (AFM-SECM). *Electrochem. Commun.,* **2007**, *9*, 1311-1315.
[http://dx.doi.org/10.1016/j.elecom.2007.01.027]

[48]   Schulte, A.; Belger, S.; Etienne, M.; Schuhmann, W. Imaging localized corrosion of NiTi shape memory alloys by means of alternating current scanning electrochemical microscopy. *Mater. Sci. Eng. A,* **2002**, *378*, 523-526.
[http://dx.doi.org/10.1016/j.msea.2003.10.354]

[49]   Diakowski, P.M.; Baranski, A.S. Positive and negative AC impedance feedback observed above conductive substrates under SECM conditions. *Electrochim. Acta,* **2006**, *52*, 854-862.
[http://dx.doi.org/10.1016/j.electacta.2006.06.020]

[50]   Annergren, I.; Zou, F.; Thierry, D. Application of localized electrochemical techniques to study kinetics of initiation and propagation during pit growth. *Electrochim. Acta,* **1999**, *44*, 4383-4393.
[http://dx.doi.org/10.1016/S0013-4686(99)00154-1]

[51]   Ningshen, S.; Mudali, U.K.; Baldev, R. Corrosion assessment of nitric acid grade austenitic stainless steels. *Corros. Rev.,* **2009**, *51*, 493-531.

[52]   Blanc, C.; Orazem, M.E.; Pebere, N.; Tribollet, B.; Viver, V.; Wu, S. The origin of the complex character of the ohmic impedance. *Electrochim. Acta,* **2010**, *55*, 6313-6321.
[http://dx.doi.org/10.1016/j.electacta.2010.04.036]

[53]   Wu, S.; Orazem, M.E.; Tribollet, B.; Vivier, V. Impedance of a disk electrode with reactions involving an adsorbed intermediate: experimental and simulation analysis. *J. Electrochem. Soc.,* **2009**, *156*, C214-C221.
[http://dx.doi.org/10.1149/1.3123193]

[54]   Schneider, L.A.; Kramer, D.; Wokaun, A.; Scherer, G.G. Spatially resolved characterization of PEFCs using simultaneously neutron radiography and locally resolved impedance spectroscopy. *Electrochem. Commun.,* **2005**, *7*, 1393-1397.

[http://dx.doi.org/10.1016/j.elecom.2005.09.017]

[55]   Lima-Neto, S.; Farias, J.P.; Herculano, L.F.; de Miranda, H.C.; Arujo, W.S.; Jorcin, J.B.; Pebere, N. Determination of the sensitized zone extension in welded ISI 304 stainless steel using non-destructive electrochemical techniques. *Corros. Sci.,* **2008**, *50*, 1149-1155.
[http://dx.doi.org/10.1016/j.corsci.2007.07.014]

[56]   Baril, G.; Blanc, C ; Kedam, M.; Pebere, N. Local electrochemical impedance spectroscopy applied to the corrosion behavior of an AZ91 magnesium alloy. *J. Electrochem. Soc.,* **2003**, *150*, B488-B493.
[http://dx.doi.org/10.1149/1.1602080]

[57]   Mohan, S.; Sivakumar, N.S.; Ajay, A.V. Senthil Saravanan M.S., B.R. Corrosion behaviour of ZrO_2-TiO_2 nano composte coating on stainless steel under simulated marine environment. *Materials Today: Proceedings,* **2020**, *27*, 2492-2497.

[58]   Titz, J.; Wanger, G.H.; Spahn, H.; Ebert, M.; Juttner, K.; Lorenz, W.J. Characterization of organic coatings on metal substrates by electrochemical impedance spectroscopy. *Corrosion,* **1990**, *46*, 221-229.
[http://dx.doi.org/10.5006/1.3585095]

[59]   Mansfeld, F. The use of electrochemical impedance spectroscopy (EIS) for the study of corrosion protection by polymer coatings—a review. *J. Appl. Electrochem.,* **1995**, *25*, 187-202.
[http://dx.doi.org/10.1007/BF00262955]

[60]   Skale, S.; Dolecek, V.; Slemnik, M. Substitution of the constant phase element by Warburg impedance for protective coatings. *Corros. Sci.,* **2007**, *49*, 1045-1055.
[http://dx.doi.org/10.1016/j.corsci.2006.06.027]

[61]   Cano, E.; Laufente, D.; Bastidas, D.M. Use of EIS for the evaluation of the protective properties of coatings for metallic cultural heritage: a review. *J. Solid State Electrochem.,* **2010**, *38*, 381-391.
[http://dx.doi.org/10.1007/s10008-009-0902-6]

[62]   Darowicki, K.; Krakowiak, S.; Slepski, P. The time dependence of pit creation impedance spectra. *Electrochem. Commun.,* **2004**, *6*, 860-866.
[http://dx.doi.org/10.1016/j.elecom.2004.06.010]

[63]   Krakowiak, S.; Darowicki, H.; Slepski, P. Impedance of metastable pitting corrosion. *J. Electroanal. Chem. (Lausanne Switz.),* **2005**, *575*, 33-38.
[http://dx.doi.org/10.1016/j.jelechem.2004.09.001]

CHAPTER 2

Anti-Corrosion Coating Mechanisms

D. Chandra Sekhar[1]**, N. Suresh Kumar**[2,*]**, K. Chandra Babu Naidu**[3]**, B. Venkata Shiva Reddy**[3,4] **and T. Anil Babu**[3]

[1] *Department of Physics, Govt. College (A), Anantapuramu-515002, A.P., India*

[2] *Department of Physics, JNTU College of Engineering Anantapur, Anantapuramu-515002, A.P., India*

[3] *Department of Physics, GITAM Deemed to be University, Bangalore - 562163, Karnataka, India*

[4] *Department of Physics, The National College, Bagepalli-561207, Karnataka, India*

Abstract: This book chapter discussed the various anticorrosive coatings, namely barrier, inhibitive, sacrificial, inorganic coatings, organic coatings, surface treatment, and importance of inhibitors to reduce the corrosion rate in metal and alloys. The various modified coatings were introduced to reduce pores, and defects observed in conventional coatings. The water permeability through the barrier coatings was also reduced using pigmentation of organic coatings.

Keywords: Anticorrosive, Hybrid inorganic-organic Coatings, Inhibitors, Inorganic and Organic Coatings, Sacrificial Coatings.

1. INTRODUCTION

In general, metals can be protected from corrosion using anticorrosive coatings categorized according to the mechanical process. The three basic anticorrosive coatings were known to be the protection of barrier, the effect of inhibitor (passivated substrate surface), and sacrificial type protection (galvanic effect). The small permeability of ions, gases, and liquids in a coating used for the substrate's surface provides a barrier layer which prevents the transportation of combative specimen into the substrate's surface. The chemical conversion layer coatings and inhibitive coatings provide the passivation of the substrate surface. The extensive application of metallic, organic, and inorganic coatings offers

* **Corresponding author N. Suresh Kumar:** Department of Physics, JNTU College of Engineering Anantapur, Anantapuramu-515002, A.P., India; Tel: +91-81211 27157; E-mail: sureshmsc6@gmail.com

N. Suresh Kumar, P. Banerjee, H. Manjunatha and K. Chandra Babu Naidu (Eds.)

sacrificial protection to protect metals against corrosion. An electrochemical metal with sacrificial corrosion can be protected if the metal is in electric contact with the substrate [1]. A few review studies on metallic coatings were reported [2, 3].

2. DIFFERENT COATING MECHANISMS

2.1. Barrier Coatings

Primer, topcoat, or intermediate types of barrier coatings were used frequently on immersed structures [4]. In general, the use of iron oxide, titanium dioxide, and lamellar aluminium in a small amount of pigmentation can illustrate the barrier coatings. The concentration of lower pigment volume introduces densified nature of cohesive coatings. At this juncture, the permeability can be lower towards combative specimens than either of the two other types of coatings [5]. The protection rate provided by a barrier coating system depends largely on the thickness of the coating system, the standard type, and nature of the binder system.

In literature, it was reported that the delamination of both less defective, and degraded barrier coatings significantly was reduced while increasing the thickness of the coating [6, 7]. Because coatings can act as impermeable films [8]. Indeed, the anticorrosive efficiency of barrier coatings was enhanced due to the use of the thickness of the same film which was built from several thin coats instead of a singular coat [9]. These coatings provided a barrier to prevent the corrosion, non-permeability of both oxygen, and water from the surroundings [10]. The protective mechanism of the barrier depends on the non-permeable ions into the coatings [11 - 13]. The humidity at the interface of substrate-coating contains a large electrical impedance due to the ionic impermeability of barrier coatings. Hence, the transfer of corrosive current from anode to cathode was decreased due to less conduction of the electrolyte solution at the substrate [3]. Cathode protective mechanism was added to the barrier coatings to prevent the damage of the substrate. The main role of the cathode protection mechanism would affect the external current to the specimen. Two distinct approaches are availed to produce external current:

- The noble specimens used as sacrificial type anodes are bonded by metals.
- Since external current source is used as a rectifier. The use of referred electrode shall control the current *via* a rectifier.

The use of sacrificial anodes can protect several offshore systems. This may even be only the protective system on submerged parts of the structures. Low-cost aluminium alloys as sacrificial anodes were used to protect the steel on the offshore structure in seawater. Zinc anodes also act as sacrificial anodes which were generally applied on offshore coated and buried structures of pipelines, whereas the low current density of aluminium sacrificial anodes offers high passivation. The magnesium sacrificial anodes acquired a large driven potential which was compared with zinc and aluminium anodes. Hence, magnesium anodes were used to a great extent for some applications in large resistivity atmosphere, like the structure of steel in soil, and in the heated water container for pure water. The main advantages of impressed current compared with sacrificial anodes were lower anode weight and lower drag forces from the sea. According to theory, the impressed current compared with sacrificial anodes shows economically beneficial while practically expresses the disruptions of the protective system of impressed cathode current, and its applications which were pervaded into current systems [14].

Another challenge is the combination of cathodic protection and coatings, which were applied to alkali specimens obtained from hydroxyl ions. These were formed with the oxygen reduction process. It indicates that the coatings in combination with cathodic protection cannot suitable choices for saponification. The good barrier coatings were produced using some binders. The use of polymers of hydrolyzed steady bonds, *e.g.,* epoxies and urethanes, produced the barrier coating systems much better than polymers containing several hydrophilic groups, *e.g.,* alkyds [15, 16]. The presence of polar groups usually increases the adhesive strength to the substrates of metals due to the polar groups. These were bind to the surface of metal-oxide through the bonds of hydrogen element [16]. Moreover, the restriction of binder choice depends on the nature of the environment where the barrier coatings were used. The well-designed barrier coatings can reduce the corrosion under the demand of specific requirements in addition to the combinations of inhibitors, immersed in fresh and burial soils, seawater, and well service in high corroded environments. The more crosslink density became a key element for the greater efficiency of barrier coatings [17, 18]. Several research studies emphasized ionic conduction process in the coatings [19 - 21], and the direct relationship between the existence of small crosslinked density in a film and the formation of under film type corrosion [22].

2.2. Sacrificial Coatings

The design of sacrificial coatings depends on the principle involved in the galvanic type of corrosion. These coatings were used to protect the metals from corrosion. The use of more electrochemical active metal/alloy in sacrificial coatings results in the strong protection of the substrate. Hence, the coatings were equipped with zinc and these were widely used for protecting the steel systems from corrosion [23]. These coatings were used as primers which can be active in the coating due to the sacrificial metal as an electrical contact with the substrate. Subsequently, the limited applications of sacrificial coatings were more responsible for the systems partially merged in water. This was happened due to the diffusion of water causing the corrosion on the sacrificial. The use of metal like zinc produces an anode type coating in the rich element zinc primers [24]. Zinc metal served as an anode initially and later turned into cathode after sacrificing itself to protect the metal. The resistivity of corrosion depends on the zinc primer transmitting galvanic current up to the conduction in the structure. This can be retained until sufficient zinc serves as the anode, the galvanic protection of metal.

In degraded zinc anode coatings, the addition of electrochemical activity results to obtain corrosive zinc specimen. These were used to close the pores between the Zn elements to a particular position where the system turns into an insulator. Additional protection indicates the barrier effect of corrosion products [25, 26]. The efficiency of sacrificial type coatings depends on the movement of the galvanically current. The formation of metallic bond lies between the independent species of sacrificed metal. Since these coatings with high pigment concentration were generally at the below the concentration of critical pigment density. Certainly, the high electrical conduction can be obtained at zinc element concentration in the range 92–95 wt % of the dried film [27, 28]. Binders with 5–8 wt % were added to the coating system and a small concentration of other specimens. For protecting the physical properties, the capacity of adhesive and cohesive forces, the effective resistance was minimized.

2.3. Inhibitive Coatings

Inhibitive coatings can be used generally used as primers. These were effective when the metal reacted with dissolved particles. In corrosive atmospheric environments like industries, the substrates can be coated with inhibitive primers. The mechanism involved in anticorrosive inhibitive coatings can become the passivated substrates which were made by a protecting layer consisting of non-soluble metal composites. This can act as a barrier to prevent the transport of combative species. In general, the inorganic salts can be used as inhibitive pigments, and these were found to be partially soluble in water. Since widely used cations were the phosphates [29, 30], most of the studies reported that chromates [31 - 33], molybdates [31, 32, 34, 35], nitrates, borates [37], and silicates [36] were the cations used frequently in the case of inorganic salts. The components of the pigments were partially dissolved and carried to the substrate surface due to the permeation of coating by moisture. The dissolved ions able to react with the substrate at the substrate's surface and produce a reacted product. This allows the passivation of the substrate's surface [37]. It indicates that the pigments of inhibitors must be required sufficiently to filter from the coating. However, blistering can occur due to the high number of pigments which were soluble [38].

A barrier can be formed against water and adverse ions in the case of an inhibitive coating. Consequently, an appropriate number of inhibitors released simultaneously. So, these two preconditions were in fact, antagonistic. Hence, the equilibrium state can be reached between the properties of barrier and the effective nature of the inhibitor. The performance of inhibitive pigments was more dependent on the barrier properties of a coating. Predominant active nature of the barrier pigments was observed due to the low diffusion of the coating system. However, the effective nature of the pigments of inhibitor will be high realistic during the coatings with a desired amount of permeability since the soluble pigments and the transferred mass in the premises of the coating would be dominant in this aspect [39].

2.4. Inorganic Coatings

Inorganic coatings were prepared using the minerals, like quartz, earth strata, and dyes obtained from inorganic mineral. Zinc metal silicates became the most used inorganic coatings. In the zinc metal silicates coating, the pigment can be added above the concentration of critical pigment volume [40]. The polymer will be coated using pigment elements, but not all and some vacancies seemed between elements which are not filled with polymer. It indicates that these coatings are porous. The large, pigmented zinc metal silicates were the good protective coatings to prevent corrosion in case of special coating applied.

The huge pigmentation of zinc metal silicates contains a huge risk of cracks. In the anticorrosive coating, primers like metallic zinc silicate or zinc metallic epoxies were extensively used [41 - 43]. The sol-gel method based inorganic coatings depend on the diffusion of inorganic metallic salts with particles. Extensive studies reported on the sol-gel method of producing zinc oxide, silicon oxide, and TiO_2–SiO_2 coatings can protect the metals from corrosion [44 - 47]. In the sol-gel coatings, inorganic material particles usually offer a good barrier to prevent the diffusion of combative species [48]. Since it can be severe to use coatings by the sol-gel method to reach the desired thick coating and obtain strong anticorrosive without cracking. Moreover, the brittleness of sol–gel-based coatings was used at huge operating temperatures and not applicable to high-scale systems [49].

An interesting feature was found from past research having the combination of the organic-polymeric samples. In addition, the inorganic material ceramics enabling the studies in hybrid organic-inorganic coatings can protect from corrosion [50]. The evolution of hybrid organic-inorganic ceramics produced by the sol-gel method provides the incorporation of organic polymeric particles into an inorganic matrix. Subsequently, the increase of more organic coatings, and the existence of polymeric particles can enhance the flexible mechanical and hardness of the coating [51]. The adhesion was enhanced in the case of hybrid inorganic-organic materials while comparing with perfect organic coating. Hence, inorganic materials can interact with metallic surfaces [52]. The introduction of polymeric species will reduce the porosity of the coating by seizing the pores between the inorganic species, which enhances the properties of the barrier. The extensively used hybrid inorganic-organic materials can be the modifications of organosilanes (polysiloxanes). These were produced by hydrolysis-condensation of modified organic silicates with typical alkoxide starting materials [53]. The current research on anticorrosive coating paid great attention to the modification in hybrid organic-inorganic coatings done by inorganic materials or extenders like phosphates [54].

Phosphate, and chromate material conversion coatings were of well-known inorganic conversion coatings. Phosphate inorganic coating can be formed on the metallic surface by immersing the metal in a chamber consisting of phosphoric acid and chemicals. In the immersion process, the metallic particles can be able to react with phosphoric acid and chemicals in the chamber. As a result, it forms a protective non-soluble phosphate layer. The resistivity of corrosion and adhesion strength of paints were increased by the application of phosphate coatings. The commonly used phosphate conversion coatings were $Zn_3(PO_4)_2$, $Mn_3(PO_4)_2$, and $FePO_4$. The principle of phosphate conversion coating can be divalent metallic, and phosphate ions which were placed on the surface of the metal by the precipitation method. The phosphate conversion coatings were applied by spray or immersion methods.

Metallic coatings in chromium (Cr^{6+}) solutions were subjected to the electrochemical method. Then, the chromate conversion coatings were formed on the metallic surface. Chromate conversion coatings were used to enhance the resistance of corrosion for the metallic structures and to increase the adhesive strength of painted metallic surface. Chromate conversion coatings were applied by spraying as well as by immersion methods. However, it was found to be necessary to replace the chromate conversion coating by green alternative due to the carcinogenic and toxic nature of hexavalent chromium.

Anodization was noticed to be the modification in surface performed for aluminium metallic alloys to enhance the corrosion resistivity. If aluminium and its based alloys were polarized anodically in electrolyte solutions, the aluminum oxide (Al_2O_3) films will be produced on the surface. Hence, all the anodic mechanisms form porous films of Al_2O_3. It was necessary to seize the pores with the immersed coating specimens in the hot water container. The coatings were hydrated using hot water and sealing the pores by creating $Al_2O_3.3H_2O$. Active passive was the newest approach to protect the coatings and the protection of the coating layer acting as a barrier. It inhibits the transport of corrosive species to the bottom of the metallic surface (passive). On the other hand, the use of the active approaching method will produce a passivated layer which disrupts corrosive reactions. At the interface, this passive layer served as a Schottky barrier which results in the depletion of a greater number of electrons.

2.5. Passivation: Anti-Corrosion Coating

The resistivity of polarization methods was used to determine the rate of corrosion for kinetic coating systems. The main parameter was noticed to be current density, and it was used to measure the performance of protection for kinetic system coatings. More current density seemed for the low electrochemical activity. In addition, a low voltage of both passivation and pitting may cause a large amount of corrosion due to the existed defects and pores in the coating mechanism. Therefore, the barrier layer became a failure when the electrolyte diffusion and endorsing occurred [55].

3. ANTI-CORROSION COATINGS

3.1. Polymeric Materials

Scientists reviewed [56] the polymeric materials and suggested them as the anti-corrosion coatings. In addition, few polymeric materials inhibit corrosion. In view of the invulnerable nature of the polymeric materials [57, 58], researchers [59] improved modern anti-corrosive coating using polyaniline (PANI), and an organic metal. Polyaniline is the organic metal that passes the bottom of metals by modified surface potential. The intrinsic conducting polymers (*i.e.,* ICPs) can be used as anti-corrosion coating species on the surface of the metal [60 - 67]. As the preparation of ICOs can be a very easy process at low cost, these materials became thermally stable. Moreover, the PANI was observed as the frequently used metal for the ICPs [68 - 71]. The additional benefit for the ICP materials was like a single coating layer usage. It can be necessary for multi-layer coating of conventional material samples to protect from the corrosion to the bottom of the metallic surfaces. An organic polymeric material usually consists of different functional chemical elements. The most used organic polymers were asphalt mastic, asphalt enamel, coal tar epoxy, fusion-bonded epoxy, extruded polyethylene, multilayer polyolefin coating systems, and polyurethane.

3.2. Metallic Coatings

Metals and metallic alloys were applied to another metallic material and their alloys to inhibit corrosion. A metallic coating can act as noble coatings as well as sacrificial coatings. The metal surface was coated with metallic materials by the following methods:

3.2.1. Electroplated Coatings

Electroplated coatings were defined as the adherent metallic coatings upon metal and its alloys which can be achieved using the electrodeposition technique [72]. In electroplating process, a thin coherent metallic coating will be formed on an electrode by using an electric current to minimize dissolved metallic cations. These coatings can provide protection from corrosion, corrosion resistivity, decorative, electric, and magnetic properties. The desired thickness of the coating can be achieved by the modification of few parameters like temperature, time, current density, and composition of the bath. In general, the deposition of zinc and cadmium metals on steel can act as active coatings. Besides, the steel can protect by cathodic protection. Various metallic materials coated on steel were known as noble coatings. These materials act as a barrier to prevent the corrosive agents, and to confirm that noble coating would be free of pores and flawless.

3.2.2. Electroless Metal Coatings

Electroless metal coatings were produced in the absence of an electrical current. In aqueous solution, the nickel ions treated with catalytic chemical reduction process can produce electroless nickel coatings. Herein, the prepared nickel and phosphorous boron coatings offer strong corrosion resistivity in various environments [73].

3.2.3. Hot-dip Coating

Hot-dip coating is a process where a less melting metal can be used as a protection of coating on steel in which the coated material is immersed in a molten bath. Even though the hot-dip coating method can be used over several material species, it was generally used to protection of steel. Hot-dip coating is a continuous processing method. Al and Zn metals can be generally used for coating the steel. Zn metal coated steels by hot-dip method are called galvanic steels [74].

3.2.4. Thermal Spraying

Thermal spraying is a process which includes the spraying of flames and plasma, detonation gun, the cold spray of oxyfuel with more speed. This can be applied to various coating species for corrosion resistance. Powder, rod shape, and liquid forms of materials were used in the coating. Heat treatment of the coating species was shifted to a state of plasticity or molten. These were driven by a streamed and compressive vapor onto the surface of the substrate. When these particle species collide with the metallic surface, they become flattened and produce thinner platelet and strongly adherent to the surface. In general, a spraying gun can be applied to create the desired heat for melting *via* combustion of gas vapors, plasma, or electrical arc.

3.2.5. Cladding

Cladding is the bond formation between different metals together. Two metallic sheets are rolled together. The cladding principle contains cold-rolled bond, hot-rolled bond, weld cladding, and explosive bond. For example, Ni and metallic steel sheets are hot-rolled for producing a Ni-steel composite sheet. Likewise, strong aluminium metallic alloys are cladded with pure aluminium metal to have a suitable barrier should prevent the corrosion.

4.2.6. Vapor Deposited Coatings

Vapor deposited coatings consist of a chamber where the vacuum is higher. They classified into two categories of vapor deposition such as chemical vapor deposition (CVD) and physical vapor deposition (PVD). Generally, the electrical heat treatment is applied to coat material, which is vaporized. These vapors are deposited on the specified protective component. The main vapor deposition methods are evaporation, sputtering, ion plating and CVD. Usually, the materials are deposited in the form of independent atoms or molecules. The vapor deposition is used for thin coating offers coatings with high density and pores less. Hence, the limited applications of VD are used for the protection from corrosion since this method is economically high expensive in comparison with other methods.

3.2.7. Ion Implantation and Laser Processing

Ion implantation and laser processing methods are two modified surface mechanisms which are produced using particle beam or higher energy. The corrosion and wear types of combative issue appear on the surface. Hence, it is necessary to modify the surface and it can be achieved using high-energy ion beams. The use of ion implantation made it easy to increase the passivated features or produce a noble species. Generally, application of ion implantation is used in the semiconductor industry. Lasers with output power in the range of 0.5-10 kW can be applied to the metallic surface. For tailoring its surface properties, they are intact with its bulk properties. The laser process can create corrosion resistivity for layers of the surface. Surface alloying, melted surface, and transformation hardening can be achieved by laser processing.

3.3. The Organic Coating System

Indeed, organic coatings are the barriers between the bottom of the base metal and the corrosive region. This coating provides a durable structure, atmosphere resistance, humidity, abrasion, toughness, and aesthetic existence. The efficiency of the organic coating relies on the coating mechanical properties, inhibitive concentration [75, 76], treatment of the metallic surface [77], the adhesive strength between a coating to the bottom of the metal base [78], and other species that protect the substrate from the corrosion. The formulated coating generally consists of solvents, resins (binders), pigments, fillers, and additive species. As the coating is applied to the bottom of the metal, it establishes a uniform coating which inhibits cracks during internal stress, physical ageing, and water permeation. Anticorrosive coatings contain less permeability, strong resistivity of corrosion, and durability.

Organic coatings are categorized based on the chemical composition of the resin or binder. The binder is suspended or dissolved in the liquid. The conventional binders (used to design individual component organic coatings) are the compositions of vinyl and acrylic, chloridized rubber, alkyds, alternation of alkyds-silicon, phenolic alkyd, amino- changes of alkyd, and epoxy ester. The combinations of phenolic and polyurethanes will produce double component organic coatings. Coating, barrier properties and diffusion of water depending on the chemical species in the pigment, concentration of pigment, and concentration of critical volume. The pigment species can offer the protection of binder against ultraviolet radiation in addition to the color and opacity. Binders used to control the properties of coating along with durability, flexible nature, curing duration, efficiency, and adhesive strength [79].

Several functions were performed by organic solvents. Since binders are soluble in solvents, they can control coating and evaporation to obtain the film. In addition, they can affect film adhesive strength and coating duration, other additives. Herein, filler species offer uniform coating and improving coating, inhibit the diffusion of H_2O and O_2. Metallic surface prior painted analysis like chromium and phosphate conversion coatings is used to improve the adhesive strength of the organic coating. Due to the strong surface adhesion and inhibitive properties of a primer coat was applied initially before the topcoat. More than one coat establishes pleasant color and opacity, good mechanical properties, and good barriers properties (inhibit the permeability of H_2O and O_2 through the interface among the bottom of the meal and coating).

3.4. Pigments

Pigments usually provide color and opacity. The advantages of pigmentation can prevent water diffusion and protecting the cured binder against ultraviolet radiation. Pigments also act as volume fillers to reduce the cost of coating. Examples for pigment species are titanium dioxide, iron oxides, and carbon. The size of pigments is in the range between 0.01 and 10 μm [80]. The protection primer coatings were done using the pigments: Chemical resistance of pigment species produces barrier coatings for alkaline, and acidic atmosphere, zinc-rich type pigments of sacrificial can offer galvanized protection in the pH range 5-10, and (iii) effective pigments like chromate species act as inhibitors in alkyd primers [81 - 83].

Most of the studies were operated to understand the properties of pigment. Moreover, they played a role in formulating paintings, production, and application. The out comings obtained for various coating properties such as internal stress, diffusion of water, glass transition temperature, physical ageing, and formation of films are suddenly altered by the chemical components of pigment, pigment level, size of the particle, and dispersed pigment [84 - 89]. Anticorrosion pigments widely applied in the industrial of paints are integrated into primary paints to facilitate the properties of anticorrosion. The coatings protect against salt corrosion, moisture, O_2, and corrosion gases like sulphur dioxide and sulphur trioxide [90, 91]. The application of pigments shows the effect on properties of organic coating, cured and formation of films [92], coefficient of thermal expansion [93 - 95], properties [96 - 98], glass transition temperature [99, 100], physical ageing [101 - 103], water diffusion [104 - 106], and enhanced stress [107 - 109]. The effective pigmentation on the properties of organic coatings was reported in the review literature [110].

3.5. Solvents, Additives and Fillers

Binders are dissolved in suitable organic solvents and further applied to evaporate from the coating. Additive species in low concentration are applied as inhibitors (Cu_2 compounds), surface-dried additive materials (Mn and Co naphthenate), and barrier-creation additive species to protects the binder from ultraviolet radiation. Mica and talc fillers reduce the diffusion of O_2 and H_2O in the coating.

4. APPLICATIONS

Organic coatings are widely used in applications like painting, powder coating, E-coating, polymerized, and sol-gel based coating [111]. The selection of primer is the primary step in painting. Powder coatings are classified into two types. First one is thermoplastic, and later one is thermosetting. When the heat in the flame is applied to thermoplastic powders, they get melted and finally, they remain with the same chemical structure after treatment of cooling and solidifying process. Thermosetting powder coatings are also melted for the heat treatment since they are chemically cross-linked within themselves or with the other reactive materials. After the curing process, the coating gained a chemical structure other than the basic resin. Thermoset powders can also be used by electrostatic powder spraying method to produce thinner films with good appearance than some thermoplastic powder coatings. Two-step methods are involved in E-coating or electrophoretic deposition (EDP). During the first step, electrophoresis is applied to transport the charged elements to an electrode (substrate). In the second step, particle deposition is used to produce the compacted film [112]. The multistep process was involved in sol-gel painting. A solid-state is formed during the first step through the gel transition of a colloid phase. Drying treatment used to obtain "dry gel," or xerogel is the second step. Rest of the steps are applied to enhance the density, stable phase, inserting, or removing unnecessary residues of organics.

Many organic compounds are inhibitors to corrosion of steel. More numbers of polar substances tend to adsorb the high energy surfaces. Amines are extensively used for this purpose. Cleaned steel was wrapped in paper introduced with a volatile amine, or the amine salt of a poor acid which is protected against corrosion. Amines are also used in boiler water to decrease corrosion. They behave as inhibitors used to neutralize acids since amines are strongly adsorbed on the surface of the steel by hydrogen bonding or salt formation, with acidic sites on the surface of the steel. This adsorbed layer act as a barrier to inhibit oxygen and water from reaching the surface of the steel. However, this mechanism functions only in the presence of an aqueous medium containing the inhibitor.

4.1. Smart Coatings

The developed smart coatings can either minimize the corrosive process or estimate the commencement of corrosion in a particular applied material. The smart coating can be defined as the several coatings to modify properties of species in responding to environment catalyst and prefer in a reversible fashion. An anti-corrosion smart coating is provoked suddenly after the commencement of corrosion in the application. The scientists reported that the substrate Al-2024 was used for a smart coating. It was produced by the sol-gel method to obtain the passivated corrosive protecting barrier. The coating contains ZrO_2, Si, and a binder. The combinations of polyelectrolytes applied as the beginning of corrosion sensor depend on the pH of the environment and in fluids with neutralized pH. The polyelectrolyte-inhibitor combination was noticed as a stable phase. A polyelectrolyte combination was steady in acid pH fluids applied. But still, this combination discharges their inhibitors in a base pH environment. The smart coating shows the strongest adhesive to the bottom of the Al-substrate, and it enables reversed the galvanized principle at the commencement of corrosive sites.

4.2. Recent Advances in Protective Coatings

The organic/polymeric coatings are used in applications to protect the metal and metallic alloys. Polymer coating system consists of a conversion layer, primer, and topcoat. These conversion layers are used to increase the performance of corrosion resistance and further to have the strong adhesion of organic coatings. Phosphate and chromate conversion coatings are the applied techniques to attain a conversion layer, increasing the adhesive strength of organic coatings for both ferrous and non-ferrous alloys [113]. Even though the high oxidization of Cr (VI) results in various hazard complications to human livings, and it is carcinogenic. The major drawback of chromates and phosphate type corrosion inhibitors are toxic in nature. However, there is necessary to establish a sustainable environment for the pretreatment of the surface. Sol-gel coatings are alternatives to previous mentioned toxic inhibitors for the protection of surface from the corrosion. A sol-gel with dip coating is an elementary method containing the extraction of a substrate product from a sol (fluid). The gravitational drain is required to obtain a thinner solid film by the deposition of the product. Further, evaporation of the solvent and then the condensation reaction occurred. In comparison with other thinner solid film deposition methods like PVD, CVD, electrochemical deposition, and sputtering, the sol-gel dip-coating was the best method. It requires a very small number of devices, and it is a low-cost method less.

The main advantages of sol-gel coatings are the highest corrosion resistance, non-toxic, compatible with the environment, strong adhesive nature to the metal substrate and top organic coating. The sol-gel technique is applied to produce the protection coatings like organic and inorganic. In addition, the sol-gel process requires low reaction temperatures, and it made it easy to introduce the organic composition into the inorganic species. This made easy for the synthesis of a modern category of hybrid coatings containing the organic and inorganic materials. Moreover, organic coatings are usually thicker, free of pores, and more density. However, the barrier coatings offer corrosion resistance sometimes at incidence. The failure of the barrier layer causes the permeability of corrosive species to the metallic surface resulting in the beginning of corrosion. It is necessary to introduce the corrosion inhibitive materials in the barrier coatings to enhance its anticorrosive nature. Corrosion protection offered by active corrosive inhibitors is called as effective corrosion protection. In the case of active corrosion protection techniques, the corrosion inhibitive materials are introduced into the anticorrosive barrier layers to minimize the rate of corrosion during onset degradation of the passivated barrier layer. There are two approaches for the incorporation of corrosive inhibitors into the coatings: (i) Introducing the inhibitors into the indirect coating approach (ii) inhibitors are encapsulated in a few chambers and further incorporated into the coating. The direct introduction of inhibitors shows a negative result about the stable coating of the sol-gel method, and the function of the barrier starts degrading. When coming to the second approaching, it is preferable for the inhibitors encapsulated in the chamber prior to mix coating.

Recent improvements in surface engineering and nanotechnology paid attention to synthesizing novel category of coatings. This offers both protection of barrier protection and functional activities. Nano or microchambers are sustained in releasing properties and they can be applied in a new category of self-healing coatings. The inhibitors are released quickly from the nano-chambers during the modifications of environments. These inhibitors obstruct activities of corrosion. Unlike a conventional coating, this self-healing coating contains several benefits for preventing corrosion. This method allows indirect contact between the inhibitor particles and barrier coatings. Further, it reduces the negative results of the integrity of inhibitor species on the coating. Electroactive conducting polymers are used to inhibit corrosion. Hence, they have a uniqueness in electrical conductivity and made them for their uses in several applications. This kind of coatings can offer high performance for the protection of soft steel. The coatings with polyaniline capable of protecting the steel in both acid and neutral conditions with the regions are exposed to re-passivation.

The self-heal polymers are encouraged by biology component systems where the catastrophe can be triggered automatically responding to the healing process. The self-heal is attained by introducing microcapsules which contain the functional components (healed species and catalyst) within a polymer matrix before formation. Coatings based on the electric current originating from the beginning of corrosion to launch self-heal should be the future scope of anticorrosion coatings. Self-healing techniques are new ones, and few results were obtained initially. They indicated that the applied technique inhibits the corrosion, and various challenges are seen.

CONCLUSION

Corrosion of materials is a huge drawback to enhance the economy and causes big damage to our human life. So that, protection from corrosion is a great task for humanity. Prevention of corrosion using anticorrosive coatings played a versatile role in the control of this threat. The formulated inorganic and organic coatings are in good adhesive nature existed between substrate and bottom of the metallic surface. Sol-gel coatings provide the barrier layers with effective functionalism. This is the best alternatives for both chromate and phosphate conversion coatings as conversion layers since sol-gel coatings are less expensive. Even though it is necessary to adopt the recent progress in nanotechnology for enhancing the properties of coatings, they should express multifunction activities.

CONSENT FOR PUBLICATION

Not applicable.

CONFLICT OF INTEREST

The author declares no conflict of interest, financial or otherwise.

ACKNOWLEDGEMENTS

Declared none.

REFERENCES

[1]   Kjernsmo, D.; Kleven, K.; Scheie, J. *Corrosion Protection*; Boarding A/S: Copenhagen, **2003**.

[2]   Choy, K.L. Chemical Vapour Deposition Coatings. *Prog. Mater. Sci.,* **2001**, 48-57.

[3]   Wilcox, G.D.; Gabe, D.R. Electrodeposited Zinc Alloy Coatings. *Corros. Sci.,* **1993**, 35-1251.
        [http://dx.doi.org/10.1016/0010-938X(93)90345-H]

[4]   Hare, C. Barrier Coatings. *J. Protect. Coat. Linings,* **1989**, 6-59.

[5]   Hare, C. *Anti-Corrosive and Barrier and Inhibitive Primers*; Federation of Societies of Coatings Technology: Philadelphia, **1979**.

[6]   Steinsmo, U.; Skar., J.I. Factors Influencing the Rate of Cathodic Disbonding of Coatings. *Corros. Sci.,* **1994**, 50-934.

[7]   Keane, J.D.; Wettach, W.; Bosh, C. *Minimum Paint Thickness for Economical Protection of Hot-Rolled Steel Against Corrosion*; J. Paint Technol, **1969**, pp. 41-372.

[8]   Sørensen, P.A.; Kiil, S.; Dam-Johansen, K.; Weinell, C.E. Influence of Substrate Topography on Cathodic Delamination of Anticorrosive Coatings. *Prog. Org. Coat.,* **2009**, *64*(2-3), 142-149.
       [http://dx.doi.org/10.1016/j.porgcoat.2008.08.027]

[9]   US Navy, Engineering and Design: Painting. *New Construction and Maintenance, EM,* **1995**, (2), 1110-3400.

[10]   Thomas, N.L. The Barrier Properties of Paint Coatings. *Prog. Org. Coat.,* **1991**, 19-101.

[11]   Dickie, R.A.; Smith, A.G. How Paint Arrests Rust. *Chemtech,* **1980**, 10-31.

[12]   Bacon, C.R.; Smith, J.J.; Rugg, F.G. Electrolytic Resistance in Evaluating Protective Merit of Coatings on Metals. *Ind. Eng. Chem.,* **1948**, 40-161.
       [http://dx.doi.org/10.1021/ie50457a041]

[13]   Kittelberger, W.W.; Elm, A.C. Diffusion of Chloride through Various Paint Systems. *Ind. Eng. Chem. Res.,* **1952**, 44-326.

[14]   Munro, J.I.; Segal., S. Cathodic Protection of Ice Shields on the Norththuberland Strait Confederation Bridge. *Mater. Perform.,* **1998**, 37-362.

[15]   *Lambourne, R, Strivnes, TA, Paint and Surface Coatings— Theory and Practice*; Woodhead: Cambridge, **1999**.

[16]   Rouw, A.C. Model Epoxy Powder Coatings and their Adhesion to Steel. *Prog. Org. Coat.,* **1998**, 34-181.
       [http://dx.doi.org/10.1016/S0300-9440(98)00018-6]

[17]   Kinsella, E.M.; Mayne, J.E.O. Ionic Conduction in Polymer Films. I. Influence of Electrolyte on Resistance. *Br. Polym. J.,* **1969**, 1-173.
       [http://dx.doi.org/10.1002/pi.4980010405]

[18]   Funke, W Towards Environmentally Acceptable Corrosion Protection by Organic Coating Problems and Realization. *J. Coat. Technol,* **1983**, 55-31.

[19]   Mayne, J.E.O.; Scantlebury, J.D. Ionic Conduction in Polymer Films. II. Inhomogeneous Structure of Varnish Films. *Br. Polym. J.,* **1970**, 6-240.
       [http://dx.doi.org/10.1002/pi.4980020407]

[20]   Ritter, J.J.; Rodriguez, M.J. Corrosion Phenomena for Iron Covered with a Cellulose Nitrate Coating. *Corrosion,* **1982**, 38-223.
       [http://dx.doi.org/10.5006/1.3593869]

[21]   Kinsella, E.M.; Mayne, J.E.O.; Scantlebury, J.D. Ionic Conduction in Polymer Films. III. Influence of Temperature on Water Absorption. *Br. Polym. J.,* **1971**, 3-41.
       [http://dx.doi.org/10.1002/pi.4980030107]

[22]   Mayne, JEO; Mills, DJ The Effect of the Substrate on the Electrical Resistance of Polymer Films. *J. Oil Color Chem. Assoc,* **1975**, 58-155.

[23]   Vilche, JR; Bucharsky, EC; Guidice, C *Application of EIS and SEM to Evaluate the Influence of Pigment Shape and Content in ZRP Formulation on the Corrosion Prevention of Naval Steel.,* **2002**.
       [http://dx.doi.org/10.1016/S0010-938X(01)00144-5]

[24]   Hare, C; Steele, M; Collins, SP Zinc loadings, cathodic protection, and post-cathodic protective mechanisms in organic zinc-rich metal primers. *J. Protect. Coat. Linings,* **2001**.

[25]   Feliu, S; Barajas, R; Bastidas, JM; Morcillo, M. Mechanism of Cathodic Protection of Zinc-Rich

Paints by Electrochemical Impedance Spectroscopy. 1. Galvanic Stage. *J. Coat. Technol,* **1989**, 61-63.

[26]   Feliu, S; Barajas, R; Bastidas, JM; Morcillo, M. Mechanism of Cathodic Protection of Zinc-Rich Paints by Electrochemical Impedance Spectroscopy. Barrier Stage. *J. Coat. Technol,* **1989**, 61-71.

[27]   Svoboda, M. *Proceedings of the XXXI International Conference on KNH,* **2000**, p. 5.

[28]   Ruf, J. *Korrosion Schutz durch Lacke und Pigmente*; Verlag W. A. Colomb, **2000**.

[29]   Cohen, M. The Breakdown and Repair of Inhibitive Films in Neutral Solution. *Corrosion,* **1976**, 32-12.
[http://dx.doi.org/10.5006/0010-9312-32.12.461]

[30]   Romagnoli, R.; Vetere, V.F. Heterogeneous Reaction Between Steel and Zinc Phosphate. *Corrosion,* **1995**, 51-116.
[http://dx.doi.org/10.5006/1.3293583]

[31]   Meng, Q.; Ramgopal, T.; Frankel, G.S. The Influence of Inbibitor Ions on Dissolution Kinetics of Al and Mg Using the Artificial Crevice Technique. *Electrochem. Solid-State Lett.,* **2002**, *5*, B1.
[http://dx.doi.org/10.1149/1.1429542]

[32]   Rafey, S.A.M.; Abd El Rehim, S.S. Inhibition of Chloride Pitting Corrosion of Tin in Alkaline and Near Neutral Medium by Some Inorganic Anions. *Electrochim. Acta,* **1996**, 42-667.

[33]   Schmucki, P.; Virtanen, S.; Isaacs, H.S.; Ryan, M.P.; Davenport, A.J. Bo¨ hni, H, Stenberg, T, "Electrochemical Behavior of Cr2O3/Fe2O3 Artificial Passive Films Studied by *In Situ* XANES. *J. Electrochem. Soc.,* **1998**, 145-791.
[http://dx.doi.org/10.1149/1.1838347]

[34]   Sakashita, M.; Sato, N. The Effect of Molybdate Anion on the Ion-Selectivity of Hydous Ferric Oxide Films in Chloride Solutions. *Corros. Sci.,* **1977**, 17-473.
[http://dx.doi.org/10.1016/0010-938X(77)90003-8]

[35]   Buchler, M.; Schmucki, P. Bo¨ hni, H, "Iron Passivity in Borate Buffer. *J. Electrochem. Soc.,* **1998**, 145-609.
[http://dx.doi.org/10.1149/1.1838311]

[36]   Sinko, J. Challenges of Chromate Inhibitor Pigments Replacement in Organic Coatings. *Prog. Org. Coat.,* **2001**, 42-267.
[http://dx.doi.org/10.1016/S0300-9440(01)00202-8]

[37]   Rammelt, U.; Reinhard, G. Characterization of Active Pigments in Damage of Organic Coatings on Steel by Means of Electrochemical Impedance Spectroscopy. *Prog. Org. Coat.,* **1994**, 24-309.
[http://dx.doi.org/10.1016/0033-0655(94)85022-4]

[38]   Prosek, T.; Thierry, D. A Model for the Release of Chromate from Organic Coatings. *Prog. Org. Coat.,* **2004**, 49-209.
[http://dx.doi.org/10.1016/j.porgcoat.2003.09.012]

[39]   Liu, WM Efficiency of Barrier-Effect and Inhibitive Anti-Corrosion Pigments in Primers. *Mater. Corros,* **1998**, 46-576.

[40]   Mitchell, MJ; Summers, M How to Select Zinc Silicate Primers. *Protect. Coat. Eur. J,* **2001**, 12.

[41]   Mitchell, M.J. Zinc Silicate or Zinc Epoxy as the Preferred High-Performance Primer *International Corrosion Conference,* South Africa**1999**.

[42]   Undrum, H Superior Protection—Silicate and Epoxy Zinc Primers. *Surf. Coat. Aust,* **2007**, 44-14.

[43]   Guglielmi, M. Sol–Gel Coatings on Metals. *J. Sol-Gel Sci. Technol.,* **1997**, 8-443.

[44]   Ballard, R.L.; Williams, J.P.; Njus, J.M.; Kiland, B.R.; Soucek, M.D. Inorganic-Organic Hybrid Coatings with Mixed Metal Oxides. *Eur. Polym. J.,* **2001**, 37-381.
[http://dx.doi.org/10.1016/S0014-3057(00)00105-1]

[45]   Schottner, G *Hybrid Sol–Gel-Derived Polymers: Applications of Multifunctional Materials.*, **2001**.
        [http://dx.doi.org/10.1021/cm011060m]

[46]   Kasemann, R.; Schmidt, H. Coatings for Mechanical and Chemical Protection Based on Organic–Inorganic Sol–Gel Nanocomposites. *New J. Chem.,* **1994**, 18-1117.

[47]   Zheludkevich, M.L.; Serra, R.; Montemor, M.F.; Yasakau, K.A.; Salvado, I.M.M.; Ferreira, M.G.S. Nanostructured Sol–Gel Coatings Doped with Cerium Nitrate as Pre-Treatments for AA2024-T3—Corrosion Protection Performance. *Electrochim. Acta,* **2005**, 51-208.
        [http://dx.doi.org/10.1016/j.electacta.2005.04.021]

[48]   Voevodin, N.N.; Grabasch, N.T.; Soto, W.S.; Kasten, L.S.; Grant, J.T.; Arnold, F.E.; Donley, M.S. An Organically Modified Zirconate Film as a Corrosion Resistant Treatment for Aluminum 2024-T3. *Prog. Org. Coat.,* **2001**, 41-287.
        [http://dx.doi.org/10.1016/S0300-9440(01)00156-4]

[49]   Messaddeq, S.H.; Pulcinelli, S.H.; Santilli, C.V.; Guastaldi, A.C.; Messaddeq, Y. Microstructure and Corrosion Resistance of Inorganic–Organic (ZrO$_2$-PMMA) Hybrid Coating on Stainless Steel. *J. Non-Cryst. Solids,* **1999**, *2*, 47-164.
        [http://dx.doi.org/10.1016/S0022-3093(99)00058-7]

[50]   Schmidt, H.; Jonschker, G.; Goedicke, S.; Menning, M. The Sol–Gel Process as a Basic Technology for Nanoparticle-Dispersed Inorganic-Organic Composites. *J. Sol-Gel Sci. Technol.,* **2000**, 19-39.
        [http://dx.doi.org/10.1023/A:1008706003996]

[51]   Hofacker, S.; Metchel, M.; Mager, M.; Kraus, H. Sol–Gel: A New Tool for Coatings Chemistry. *Prog. Org. Coat.,* **2002**, 45-159.
        [http://dx.doi.org/10.1016/S0300-9440(02)00045-0]

[52]   Seok, SI; Kim, JH; Choi, KH; Hwang, YY Preparation of corrosion protective coatings on galvanized iron from aqueous inorganic–organic hybrid sols by Sol–Gel method. *Surf. Coat. Technol.,* **2006**, 200-3468.
        [http://dx.doi.org/10.1016/j.surfcoat.2005.01.012]

[53]   Pathak, S.S.; Khanna, A.S.; Sinha, T.J.M. Sol Gel Derived Organic–Inorganic Hybrid Coating: A New Era in Corrosion Protection of Material. *Corros. Rev.,* **2006**, 24-281.
        [http://dx.doi.org/10.1515/CORRREV.2006.24.5-6.281]

[54]   Zheludkevich, ML; Serra, R; Montemor, MF; Salvado, IMM; Ferreira, MGS *Corrosion Protective Properties of Nanostructured Sol–Gel Hybrid Coatings to AA2024-T3.,* **2006**.
        [http://dx.doi.org/10.1016/j.surfcoat.2004.09.007]

[55]   Castro, Y.; Ferrari, B.; Moreno, R.; Duran, A. Corrosion behaviour of silica hybrid coatings produced from basic catalysed particulate sols by dipping and EPD *Surface & Coatings Technology,* **2005**, 228-235.

[56]   Zhang, T Tang Current Research Status of Corrosion Resistant Coatings *Recent Patents on Corrosion Science,* **2009**, *1*, 1-5.

[57]   Lee, J. H.; Brady, B. K.; Fuss, R. L. *US20097494734,* **2009**.

[58]   Kramer, T.; Gillich, V.; Fuchs, R. *US20077182475,* **2007**.

[59]   Wessling, B. Scientific Engineering of Anti-Corrosion Coating Systems based on Organic Metals (Polyaniline). *JCSE,* **1999**, 1-15.

[60]   Ahmed, N.; MacDiarmid, A. G. Inhibition of corrosion of steels with the exploitation of conducting polymers *Synth. Met,* **1996**, *78*, 103-110.

[61]   Camalet, J. L.; Lacroix, J. C.; Aeiyach, S.; Ching, K. C.; Lacaze, P. C. Lacaze: Electrosynthesis of adherent polyaniline films on iron and mild steel in aqueous oxalic acid medium *Synth. Met,* **1998**, *93*, 33-142.

[62] Sitaram, SP; Stoffer, JO Application of conducting polymers in corrosion protection *J. Coat. Technol,* **1997**, 65-69.

[63] Martins, J. I.; Bazzaoui, M.; Reis, T. C.; Bazzaoui, E. A.; Martins, L. Electrosynthesis of homogeneous and adherent polypyrrole coatings on iron and steel electrodes by using a new electrochemical procedure *Synth. Met,* **2002**, (), 129-221-228.

[64] Tallman, D. E.; Spinks, G.; Dominis, A.; Wallace, G. G. Electroactive conducting polymers for corrosion control *J. Solid State. Electrochem.,* **2002**, 6-73-84.

[65] Iroh, J. O.; Zhu, Y.; Shah, K.; Levine, K.; Rajagopalan, R.; Uyar, T.; Donley, M.; Mantz, R.; Johnson, J.; Voevolin, N. N.; Balbyshev, V. N.; Khramov, A. N. Electrochemical synthesis. A novel technique for processing multi-functional coatings *Prog.Org. Coat,* **2003**, 47-365-375.

[66] Tan, C. K.; Blackwood, D. J. Blackwood: Corrosion protection by multi-layered conducting polymer coatings *Corr. Sci,* **2003**, 45-545-557.

[67] Zarras, P.; Anderson, N.; Webber, C.; Irvin, D. J.; Irvin, J. A.; Guenthner, A.; Stenger-Smith, J. D. Progress in using conductive polymers as corrosion-inhibiting coatings *Rad. Phys. Chem,* **2003**, 68-387-394.

[68] Genies, E. M.; Boyle, A.; Lapkowski, M.; Tsintavis, C. Polyaniline: A historical survey *Synth. Met,* **1990**, (36), 139-182.

[69] Gospodinova, N.; Terlemezyan, L. Conducting polymers prepared by oxidative polymerization, polyaniline. *Prog. Polym. Sci,* **1998**, (23), 1443-1484.

[70] Bhadra, S.; Singha, N. K.; Lee, J. H.; Khastgir, D. Progress in preparation,processing and applications of polyaniline *Prog. Polym. Sci,* **2009**, (34), 783-810.

[71] Bhadra, S.; Singha, N. K.; Khastgir, D. Polyaniline based anticorrosive and anti-molding coating *Journal of Chemical Engineering and Materials Science,* **2011**, 2(1), 1-11.

[72] Abou-Krisha, M.M.; Alshammari, A.G. Electrochemical behavior of Zn–Co–Fe alloy electrodeposited from a sulfate bath on various substrate materials *Arabian Journal of Chemistry,* **2019**, (12), 3526-3533.

[73] Fayomi, O.S.I.; Akande, I.G. A Sode1 Corrosion Prevention of Metals *via* Electroless Nickel Coating: A review. *J. Phys. Conf. Ser.,* **2019**, *1378*, 022063.
[http://dx.doi.org/10.1088/1742-6596/1378/2/022063]

[74] Elewa, R.E.; Afolalu, S.A.; Fayomi, O.S.I. protective impact of molten zinc coating Sheets in contaminated environment-review. *J. Phys. Conf. Ser.,* **2019**, *1378*, 022071.
[http://dx.doi.org/10.1088/1742-6596/1378/2/022071]

[75] Yang, L.H.; Liu, F.C.; Han, E.H. Effects of P/B on the properties of anticorrosive coatings with different particle size. *Prog. Org. Coat.,* **2005**, *53*, 91-98.
[http://dx.doi.org/10.1016/j.porgcoat.2005.01.003]

[76] Popa, M.V.; Drob, P.; Vasilescu, E.; Mirza-Rosca, J.C.; Santana-Lopez, A.; Vasilescu, C.; Drob, S.I. The pigment influence on the anticorrosive performance of some alkyd films. *Mater. Chem. Phys.,* **2006**, *100*, 296-303.
[http://dx.doi.org/10.1016/j.matchemphys.2006.01.002]

[77] Grundmeier, G.; Rossenbeck, B.; Roschmann, K.J.; Ebbinghaus, P.; Stratmann, M. Corrosion protection of Zn-phosphate containing water borne dispersion coatings. *Corros. Sci.,* **2006**, *48*, 3716-3730.
[http://dx.doi.org/10.1016/j.corsci.2006.01.007]

[78] Kalendova, A.; Vesely, D.; Kalenda, P. Pigments with Ti^{4+}-Zn^{2+}, Ca^{2+}, Sr^{2+}, Mg^{2+}-based on mixed metal oxides with spinel and perovskite structures for organic coatings. *Pigm. Resin Technol.,* **2006**, *36*, 3-17.
[http://dx.doi.org/10.1108/03699420710718715]

[79]   Bortak, T.N. *Guide to Protective Coatings: Inspection and Maintenance*; United States Department of the Interior Bureau of Reclamation Technical Service Center: Washington, DC, **2002**.

[80]   Schweitzer, P.A. *Corrosion of Linings and Coatings*; Taylor and Francis Group: New York, **2007**.

[81]   Schiek, R.C. Lead chromate pigments. In: *Pigment Handbook*; Patton, T.C., Ed.; Willey Interscience: New York, **1973**; Vol. I, pp. 357-370.

[82]   Herrmann, E. Inorganic colored pigments. In: *Characterization of Pigment Surfaces*; Parfitt, G.D.; Sing, K.S.W., Eds.; Academic Press: London, **1976**; pp. 209-229.

[83]   Merchak, P. Colored organic pigments In: *The Gardner Sward Handbook: Paint and Coating Testing Manual*, 14[th]ed; West Conshohocken, PA : 100 Barr Harbor Drive , **1995** ; XVII, pp. 190-208.

[84]   Stieg, F.B. Influence of PVC [pigment volume concentration] on paint properties. *Prog. Org. Coat.,* **1973**, *1*, 351-373.
[http://dx.doi.org/10.1016/0300-9440(73)85003-9]

[85]   Toussaint, A. Influence of pigmentation on the mechanical properties of paint films. *Prog. Org. Coat.,* **1974**, *2*, 237-267.
[http://dx.doi.org/10.1016/0300-9440(74)80004-4]

[86]   Perera, D.Y.; Vandan, E.D. Internal stress in pigmented thermoplastic coatings. *J. Coatings Technology,* **1981**, *53*, 40-45.

[87]   Perera, D.Y.; Vandan, E.D. Effect of pigment on internal stress in latex coatings *J. Coating Technol,* **1984**, 56-47-53.

[88]   Al-Turaif, A.; Lepoutre, P. Evolution of surface structure and chemistry of pigmented coatings during drying. *Prog. Org. Coat.,* **2000**, *38*, 43-52.
[http://dx.doi.org/10.1016/S0300-9440(99)00085-5]

[89]   Al-Turaif, A.; Lepoutre, P. Effect of sintering on surface chemistry and surface energy of pigmented latex coatings. *J. Appl. Polym. Sci.,* **2001**, *82*, 968-975.
[http://dx.doi.org/10.1002/app.1930]

[90]   Rosi, S.; Fedel, M.; Deflorian, F.; Zanol, S. Influence of different color pigments on the properties of powder deposited organic coatings. *Mater. Des.,* **2013**, *50*, 332-341.
[http://dx.doi.org/10.1016/j.matdes.2013.03.007]

[91]   Veleva, L. Protective coatings and inorganic anti-corrosion pigments In: *The Gardner Sward Handbook: Paint and Coating Testing Manual* , 14th ed; K., Joseph, Ed.; West Conshohocken: 100 Barr Harbor Drive, **1995**; pp. 238-255.

[92]   Prime, R.B. Thermosets, in In: *Thermal Characterization of Polymeric Materials* ; K., Joseph, Ed.; Academic Press: New York, **1997**; pp. 1673-1698.

[93]   Nielsen, L.E. *Mechanical Properties of Polymers and Composites*; Marcel Dekker: New York, **1974**.

[94]   Sato, K. The mechanical properties of filled polymers. *Prog. Org. Coat.,* **1976**, *4*, 271-302.
[http://dx.doi.org/10.1016/0300-9440(76)80010-0]

[95]   Zosel, A. Mechanical behavior of coating films. *Prog. Org. Coat.,* **1980**, *8*, 47-79.
[http://dx.doi.org/10.1016/0300-9440(80)80004-X]

[96]   Narkis, M. Crazing in glassy polymers. Polymer-glass bead composites. *Polym. Eng. Sci.,* **1975**, *15*, 316-320.
[http://dx.doi.org/10.1002/pen.760150414]

[97]   Narkis, M. Size distribution of suspension-polymerized unsaturated polyester beads. *J. Appl. Polym. Sci.,* **1979**, *23*, 2043-2048.
[http://dx.doi.org/10.1002/app.1979.070230714]

[98]   Quemada, D. Rheology of concentrated disperse systems. III. General features of the proposed non-

Newtonian model. Comparison with experimental data. *Rheol. Acta,* **1978**, *17*, 643-653.
[http://dx.doi.org/10.1007/BF01522037]

[99]   Bajaj, P.; Jha, N.K.; Kumar, A. Effect of coupling agents on thermal and electrical properties of mica/epoxy composites *J. Appl. Polym. Sci,* **1995**, 56-1339-1347.

[100]   Chartoff, R.P. Thermoplastic polymers In: *Thermal Characterization of Polymeric Materials*; E.A, Turi, Ed.; Academic Press: New York, **1997**.

[101]   Simon, S.L.; McKenna, G.B. Development of isotropic residual stresses during thermoset cure: effects of cure history and resin properties *Proc. Am. Soc,* **2000**, 15-127-137.

[102]   Simon, S.L.; McKenna, G.B. Quantitative analysis of errors in TMDSC in the glass transition region *Thermochim. Acta,* **2000**, 348-77-89.

[103]   Perera, D.Y. Physical aging of organic coatings. *Prog. Org. Coat.,* **2003**, *47*, 61-76.
[http://dx.doi.org/10.1016/S0300-9440(03)00037-7]

[104]   Perera, D.Y. Water transport in organic coatings. *Prog. Org. Coat.,* **1973**, *2*, 57-80.
[http://dx.doi.org/10.1016/0300-9440(73)80016-5]

[105]   Crank, J. A theoretical investigation of the influence of molecular relaxation and internal stress on diffusion in polymers. *J. Polym. Sci., Polym. Phys. Ed.,* **1953**, *11*, 151-168.

[106]   Sangaj, N.S.; Malshe, V.C. Permeability of polymers in protective organic coatings *Prog. Org. Coat.,* **2004**, 50-28-39.

[107]   Perera, D.Y. Stress phenomena in organic coatings In: *The Gardner Sward Handbook: Paint and Coating Testing Manual,* 14th ed.; Joseph, K., Ed.; West Conshohocken, PA: 100 Barr Harbor Drive, **1995**; pp. 585-602.

[108]   Perera, D.Y. Internal stress and film formation in emulsion paints. *J. Oil Colour Chem. As.,* **1985**, *68*, 275-281.

[109]   Perera, D.Y. Internal stress in latex coatings. *J. Coatings Technology,* **1984**, *56*, 111-118.

[110]   Perera, D.Y. Effect of pigmentation on organic coating characteristics *Prog. Org. Coat,* **2004**, 50-24--262.

[111]   Khanna, A.S. *High-Performance Organic Coatings*; Woodhead Publishing Limited: Cambridge, England, **2008**.
[http://dx.doi.org/10.1533/9781845694739]

[112]   Schweitzer, P.A. *Corrosion of Linings and Coatings*; Taylor and Francis Group: New York, **2007**.

[113]   Hernandez-Alvarado Laura, A.; Luis, S.; Sandra, L.; Rodriguez, R. *Evaluation of Corrosion Behavior of Galvanized Steel Treated with Conventional Conversion Coatings and a Chromate-Free Organic Inhibitor*; Hindawi Publishing Corporation International Journal of Corrosion Volume, **2012**, pp. 368130-368138.

CHAPTER 3

Corrosion in Electronics

U. Naresh[1], N. Suresh Kumar[2,*], K. Chandra Babu Naidu[3], B. Venkata Shiva Reddy[3,4], A. Manohar[5], M. Ajay Kumar[6] and T. Anil Babu[3]

[1] *Department of Physics, BIT Institute of Technology, Hindupur, A.P., India*

[2] *Department of Physics, JNTU College of Engineering Anantapur, Anantapuramu-515002, A.P., India*

[3] *Department of Physics, GITAM Deemed to be University, Bangalore-562163, Karnataka, India*

[4] *Department of Physics, The National College, Bagepalli-561207, Karnataka, India*

[5] *Department of Materials Science and Engineering, Korea University, 145 Anam-ro Seongbuk-gu, Seoul 02841, Republic of Korea*

[6] *Department of ECE, GITAM Deemed to be University, Bangalore-562163, Karnataka, India*

Abstract: Corrosion is a defect of the material which reduces the life of the electronic device or any other devices which are fabricated with metals and alloys. Generally, using polymer coatings, the corrosion is decreased comfortably in large devices. In this chapter, we highlighted the corrosion in electronic gadgets and printed electronic circuit board. Furthermore, we discussed corrosion of electronic printing board with the possibilities, types of the corrosion as well as protective phenomena.

Keywords: Corrosion, Epoxy coatings, Galvanic Corrosion, Printed Electronic Board Assembly, Surface Mount Components.

1. INTRODUCTION

The usage of electronics and peripherals increased over the last two decades. The electronic printing device contains a greater number of transistors and its derivatives with the most convoluted circuit's arrangement for the sake of high functionality. Mobile phones are one of the examples of the high functionality and complexity in the arrangement of electronic components such as smart cameras, GPS systems, and various application support working components. In general, the usage of integrated circuits in electronic gadgets is simple, but fabrication involves an assembly of many metals and other properties like durability and

---

* **Corresponding author N. Suresh Kumar:** Department of Physics, JNTU College of Engineering, Anantapur-515002, Andhra Pradesh, India; Tel: +91-81211 27157; E-mail: sureshmsc6@gmail.com

reliable nature in all environmental aspects. On the other hand, the electronic industries provide significant applications in world technology irrespective of the environmental situation. However, the factories or companies of manufactures of electronic components could not take corrosion resistance conditions seriously. Usually, the electronic devices contain PCBA (printed electronic board assembly) as a heart. Environmental conditions like humidity (percentage of water in the atmosphere) can be provocative for the generation of a layer of water on the PCBA, which results in the corrosion generates in PCBA. This corrosion can influence the functional operation decrement extremely of PCBA [1 - 3]. The corrosion executes problems in the integrated function and multi-material predictions of PCBA in a specified electronic device. The multi-material means the device which includes a group of materials like plastics, metals, polymers, ceramic material, and complexes of different materials. Moreover, among all the materials, the main class of materials was metals and its composites. The reaction of humidity with PCBA selects those materials which were used in the PCBA of the device.

In the present days, the materials mainly used in the PCBA are noble gold, silver, tin, aluminium, nickel, copper, lead and tin. Not only these materials but also the above metal combinations are used for the layer of circuit coatings in the board. The electrical system contains both electronics (materials which control the flow of electricity) and electrical components, which include non-electrical systems such as switch, *etc* [4 - 6]. In the development of electronic devices starting from volume tube, now we use transistors which were fabricated through solid-state materials. These types of transistors make possible to electronics with comfortable size and usage in the electronic world every day. Therefore, it must be continued in the coming days. The electronic industry encountered many problems for comfortability in the origin of computers to end of kitchen devices through the development of electronics. Thus, the corrosion due to the different kinds of materials used in the electronic devices became more problematic than those materials having different states of environment [7]. In the periodic table, many elements and their complexes are used to fabricate resistors, integrated circuits, and printed board in every industrial electronic company [8]. Fig. (**1**) shows the corrosion on the circuit board.

The materials used in electronic devices face different types of corrosion. The interaction of corrosion depends on the kind of materials and metals such that some materials get corrosion slowly while some other materials get it spontaneously [9, 10]. In the group of electronic materials, the efficient material is gold. This is used at contact area in PCBA and layers of some other quickly reactive materials due to its non-reactive nature. That is why, gold is one of the costly materials and lasts for a long time. It is a well-known fact that the formation of oxide layers is defined as corrosion when reacted under certain environmental conditions. The non-reactive nature of gold restricts the oxide formation, while corrosion-efficient materials like aluminum can offer chances of oxide formation easily [10]. Generally, the circuit can be created with the help of one or two and more different metals. If the circuit is created with the help of two or more metals ions, it moves faster than the normal electronic circuit. Those different metal contacts generate salts and oxides. This process generally is referred to as galvanic corrosion. This is one of the largest problems among the corrosions of different materials and their interaction with different environmental conditions like humidity, moisture, and temperature. The galvanic corrosion generally happens only in the electrolyte, and it needs some moisture to start. This moisture can come in the water spray or relative humidity in the atmosphere. This relative humidity efficiently affects the occurrence of galvanic corrosion, and it isn't easy to control. These situations are hazardous where the electronic device is working. In addition, the temperature generated at machinery makes it dried surrounding the electronic device. Apart from the moisture, the air is also a second important factor for improving corrosion because the air contains many gases (nitrogen, sulfur components) reacted with elements and increase corrosion. Chloride, sulfur, and nitrogen ions supply the enormous corrosion of aluminium and its composites. The corrosion in big devices like ocean ships and containers may have this galvanic corrosion effect periodically due to the high dampness. The increase of temperature simply increases the rates of all these reactions [11]. The impact methods can protect from the corrosion of the electronic equipment from the small to bimachineries. The main factor in controlling corrosion is temperature. If we control the temperature variations, we can easily monitor corrosion variations.

Fig. (1). Corrosion on the circuit board.

2. FACTORS INFLUENCING CLIMATIC RELIABILITY

2.1. Humidity

The surface of printed, electronic board assembly takes corrosion mainly due to humidity interaction thereby creating a water layer. There are several ways to react to humidity with PCBA. The humidity interactions with metals or solid-water interactions can be classified into various types, such as surface interactions, condensed water interactions and water internalized into the solid. The humidity interactions can be classified as deliquescence, adsorptions, capillary condensations, absorption of water, and formation of the layer. Among all these kinds, the deliquescence, adsorption, and capillary condensation are the important interactions of PCBA. Of course, the PCBA material, architecture, surface contaminations are also the dependable parts for the corrosion. Water penetration through the absorption modification is a part of the properties of PCB laminated or bulk components. Therefore, the equipment causes internal degradation [12].

2.2. Water Absorption by the PCBA

The water absorption is caused due to penetration of wetness into PCBA and forming layers of dampness [7 - 9]. The moisture is increased before soldering. For example, at the time of storage of energy, it can cause delaminating at the later stage of manufacturing [10, 12]. When an electronic device is warmed up, water absorption can affect the thermal properties of PCB materials [13, 14]. Subsequently, the dielectric properties of laminate materials can increase the interface leakage and enhance the efficiency in multi-layer PCBA [15]. Generally, the change in capacitor's storage efficiency of electronic board and the moisture content follows a proportionality relationship. This will take moisture measurement [16, 17]. In the process of assembling of circuits with soldering of different places in the circuitry board, the interaction between the metallic elements will happen. Hence, the circuit board needs to cover a satisfactory soldering mask made by the known dielectric constant. So that it might be protected from the accumulated contaminants [18]. It does not protect the PCBA, since typically, those materials used for the circuit may have a higher diffusion coefficient.

2.3. Water Adsorption by the Hygroscopic Contaminants on the PCBA

Absorption of excess water-contaminated ions to the printed circuitry board assembly (PCBA) can be harmful to the PCBA protective surface insulation resistance (SIR). This water residue ionic activity should enhance the humidity or moisture. This parameter can meet a cut-off value of relative humidity, and the deliquescence (conversion of solid into the saturated solution it is generally referred as phase transformation, and it depends upon the atmosphere characteristics that is temperature) will form the PCBA [19 - 21]. At the stage of the relative humidity of deliquescence, the moisture may be transformed into another phase like solid by taking the support of respective temperature. If the ambient comparative humidity is a smaller amount of the deliquescence comparative humidity, the water is contaminated with the hygroscopic molecules throughout the instrument. The ionic infectivity about printed and electronic board surface may be caused due to the fabrication process or it comes from user circumstances. The etched chemicals, plating contamination on the PCBA, and polymer contaminations are a few examples of this type of absorption. The electronic service impacts heavily depending on the exposed places in PCBA. Exposed contaminants include atmospheric aerosols, and dust particles which have polluted ions. A small portion of manufacturing infects the PCBA. The soldering residues on the PCBA is happened due to water absorption and it is the cause for the electronic board corrosion. Some automated soldering processes like wave soldering or reflow soldering are considered to avoid this absorption impact. The soldering chemical reactions are used for the removal of metallic element oxides and adding the parts of elements that are causing the corrosion. The solder flux removes any contaminated parts between the corrosive and non-corrosive parts on the PCBA surface. The flux is needed to add the soldering paste as composites in the reflow soldering process, whereas the wave soldering incorporates soldering flux. It is the only chemical reaction used for the restriction of corrosion. The maintaining contact with a solid surface are controlled by the balance between the intermolecular connections of liquid to exterior and liquid to the fluid of the PCBA surface. The remained soldering flux on the PCBA is caused for corrosion in electronics [21, 22]. The dendrite structure in the PCBA is made with water adopted materials. Hence, it is easy to get corrosion. However, the cheated copper materials are used for the preparation of high SIR (tested through the standard SIR) [23]. The temperature is also an effective parameter for the removal of copper oxide circuits on PCBA. The rate of copper oxide removal is greater for adipic acid as compared to maleic acid between 140 °C- 170 °C. If the temperature comparably is less than decomposition temperature for the acid, it is made corrosive protection.

2.4. User Environment Related Contamination

The particulate contaminants and corrosive gasses are the main sources in the part of user contamination environment. The corrosive gasses like nitrogen oxide, sulfur monoxide, carbon monoxide and H_2S are the primary corrosive gasses presented in the user environment. These cases can be harmful to the PCBA parts, soldering joints and SIR [24]. Herein, sulfite and chloride, acetate, and ammonium are the main sources of particulate contamination [25]. The corrosive gasses and particulates are not deposited directly on the PCBA and electronic device working efficiency. The electronics and their working efficiency may depend on the humidity. In the PCBA, the dendrite circuit and assemblies of conductors, capacitors, resistances are encountered for all the corrosive sources [26].

3. HUMIDITY AND CONTAMINATION RELATED FAILURES

3.1. Leakage Current

The PCBA and SIR (Surface Insulation Resistance) related failure is linked together with the humidity and contamination. Generally, the PCBA is designed to work in 70% of normal humidity restrictions. The leakage current is the parameter which is arrived at due to the increase of humidity and other contamination. The stages of leakage current can exceed the levels of leak current. Further, it may exceed the critical level if humidity condensation on the PCBA surface is high within the bias voltage. The leakage current can produce corrosion generations [27]. The leakage current depends on the impact of the contamination and climate in the environment [26] to avoid the corrosion of PCBA. The conformal coatings are usually known as the protective method. The application of conformal coatings on a circuit board cleans a large amount of residue flux. The hygroscopic effect and hydrolytic processes harm the coatings and allow a high leakage current. The water absorption can create hygroscopic corrosive pollutants [28].

3.2. Electrochemical Migration

In the 19th century, metal migration was found in the presence of corrosion. The silver ions were the most important electromechanical migration. Perhaps, it is the main defect of the electronics and its operating system in the presence of a corrosive gas condensation environment. The electrochemical transformation of ions can be classified as a three-step process. That is, the first one is electro-distribution, defined as metal oxidation at the positive electrode, whereas ionic transport is defined as the transportation or diffusion of metal ions under the external electric field and the third one is the reduction of ionic metal at the dendrite nucleation site at the cathode. There are three conditions to be fulfilled for electrochemical migration to occur: i) presence of water layer supporting ion transport, ii) electrical bias, which acts to dissolve the metal surfaces as removing ions and reduces the subsequent reduction, and iii) pH conditions favorable for metal corrosive properties [29].

3.3. Formation of Anode Filament

In the case of electronic surfaces, the failure of electrochemical stability is because of the enhancement of salt from the anode and cathode interfaces. This may happen at the time of testing for the bias potential, temperature, and humidity. The CAF can be fabricated by a two-step process. That is, firstly, the dreadful conditions can be developed on the glass or substrate surface as the migration of the path of the copper conductive film, and the other one is a chemical reaction as getting water absorption of the packed and creates an electrochemical reaction on Cu conductors at the time of CAF formation. The copper corrosion soluble products will migrate through the laminated surface towards anode to cathode. While transforming conductive filament, a considerable decrease in the insulation resistance is observed. Instead of the formation electrochemical reaction or transformation, the cathode/anode filament formed at the point of the interior of PCB. The filaments contain conductive salts rather than metal ions reduced to metal. Further, the conductive salt filaments are growing from the anode to the cathode. The silver and copper corrosion on the PCBA can result in the creep corrosion mechanism since those elements are effectively influenced by the electric field in the PCBA. The creep corrosion on the top of the PCBA reduces the SIR and may lead the short circuit corrosion.

4. TYPES OF CORROSION

Corrosion failures under humidity circumstances contain: (1) leakage current failures due to surface insulation reduction, (2) electrochemical migration, (3) galvanic corrosion due to micro-galvanic cell formation, (4) creep corrosion, and corrosion caused by gaseous environment together with humidity. In the electronic printed board, different types of corrosion mechanisms may happen upon the various conditions like exposed material in the humidity, water, *etc.*, whereas the presence of low voltage is predominant among the adjacent metal surface. The types of corrosion showed different sources like lighting products and some tested products in the electronic industry.

4.1. Electrolytic Corrosion

The electrolytic corrosion mechanism arrives in some mechanical illumination products. The electronic conductors are used as electrodes in the electrolyte cell and water is used as a coolant to drop the surface temperature for the effective function of the cell. The conducting electron from an electrode is dissolved in the electrolyte solution, and it gets to an adjacent electrode. The cell solvent ionic components in the device need for the chemical reaction. Thus, there may be a chance to contaminate the cell circuit board and some other machineries, peripherals of the circuit. The subcomponents of a contaminant like residues of electric flux, cleaners, corrosion coatings and many machinery handling components that have salts, the acid may access the board. The fabricating companies are encouraged to summarize coat and insulate anodic points in the electronic board to protect metallic ions from humidity or moister contamination. If any defects happen in the electronic board coatings due to corrosion, the proper precaution of the anode electrolyte solution is taken. Hence, the electronic board must be handled with care and respective testing for the usage of the application. If a battery in the electronic system gets electrolyte corrosion, the device can get damaged.

4.2. Galvanic Corrosion Mechanism

Galvanic corrosion is different from electrolyte corrosion. It usually comes due to the two dissimilar metallic elements contamination. The soldering joints of two dissimilar elements on an electronic board can get this type of corrosion, as the electronic board is exposed to moisture or humidity. These two different elements can create a sufficient potential to the corrosion process. An electrolyte with ionic conductors is necessary to perform electrochemical reactions clearly. Thus, ionic conductors were condensed. The electrochemical migration arises due to the conductive dendrites from the neighbouring conductors. The galvanic corrosion can be happened in the IC (Integrated circuit) and connecting small & spaced copper circuits. The complicated circuit is designed by silver, copper, tin, and lead materials. It is also understood that because of some contaminations in between materials, there will be a chance of immediate galvanic corrosion.

4.3. Creep Corrosion

The copper composites are mainly the source of the creep corrosion mechanism in the printed circuit board. The oxidation process is happened, as the copper materials get corrosion. The oxidation film can form the surface of the board with the sulfide layer on the soldering mask.

5. PROTECTION OF ELECTRONIC CORROSION

5.1. Protecting Against Corrosive Failures

Low cost and high strength lightening gadgets cannot be exposed to solar radiation, humidity, and water. Anyhow, the designing of an inexpensive pico-powered lighting product that can tolerate the sunlight, dirt, humidity, rain, and frequent rough treatment is not easy for engineers. Every system is used in various conditions and goes with required applications to achieve the functionality. There is no guaranty for the product usage and warranty from the company. Herein, the primary chance is to design corrosive resistance conditions only.

5.2. Material Compatibility

The materials which are related to different products and applications must have compatibility with each other at the time of usage. Basically, the plastic materials like poly vinyl-based materials are chosen for protection from the off-gas elements which can enhance corrosion. The PVC (Poly Vinyl Carbonate) can release chlorine gases and other volatile organic mixers. These are the sources of contamination for the electrolyte formation on the surface of the dendrite circuit. The problem occurs when the new parts are attached to the circuit assembly without VOC gas injection mould. Hence, the coating and finishing are vital things for the removal of corrosion to achieve good adhesion. Sanitation is also an important factor to assure that the electronic board materials behave accordingly.

5.3. Printed Circuit Board Polish

Probably maximum of the electronic circuits are prepared with the copper element. Those electronic boards are attached with the help of soldering mask on the bottom of the surface traces. It makes preventing copper pads which are proceeding the components of the circuit. The plating of a circuit board may dissolve in the process of soldering and at the connecting points of various composites forming the galvanic corrosion. However, some points are not getting galvanic corrosion which is polished well. Whereas, in the literature, there is no evidence for the board polishing, which can protect from corrosion. The tin and lead-free hot solder finish immersion contain protective properties for the copper. The gold and nickel aluminum finishes can control the oxidation process and form strong galvanic protection. But gold metal is economically expensive. The organic solderable coatings and silver immersion is linked to creep corrosion protection in the sulfur coatings. The fabricators are supposed to encourage supporting the usage of the improvement of printed circuit board finishes and it makes increase of the corrosion resistance protection.

5.4. Conformal Coatings

Conformal coatings are to be applied on the bottom surface of the circuit board assemblage for the prevention of moisture and related corrosive environments. In corrosive protection environments, various types of conformal coatings and application processes are developed. Every method has its advantage and disadvantages.

6. TYPES OF CONFORMAL COATINGS

6.1. Acrylic Coating

This is a painting type and is the choice of the reset's corrosion. Acrylic is a material which contains transparent and mechanical stability. Hence, the acrylic coating contains better moisture resistance and ease of repair/remarks. This acrylic coating is of low cost and generally used material for the electronic industry.

6.2. Epoxy Coatings

Epoxy coating is made of epoxy resins. This type of coating is used in making plastics, painting primer, sealers, flooring coating products and various building and household applications. This can enhance abrasion and non-chemical reaction, poor in mustier rise dielectric behavior. The epoxy material is virtually not possible to remove for revising the other epoxies on a printed and electronic board. This can be energetically attacked by the application of an epoxy-based narrow piece agent.

6.3. Urethane Type

Urethane or polyurethane is a man-made crystalline composite used for manufacturing fungicides and pesticides. This one is used for the corrosive fabrication painting, which shows better chemical resistance with negligible moisture, temperature, and electronic properties. These novel properties can support the fabrication of special corrosive resistance painting.

6.4. Ultraviolet Light Curable Coatings

It is a photochemical process in which ultraviolet light (high intensity) is illuminated to dry or cure the coatings. Ultraviolet light rays formulate a liquid poly monomer with a lower percentage of photoinitiator illuminated with UV energy. These UV-cured materials enormously cure as well as provides impressive moisture, temperature resistance and chemically reactive. This type of material is only used for selective coating applications.

6.5. Silicone Type Coatings

Silicon coating is a paint which contains silicon resins to form a methyl phenyl resins mixer on the surface of the printed circuit board with very good heat resistance. This type of silicon coatings may include organic methyl phenyl resins and its modified types of epoxy-polyester, alkyd resins. Its controlled temperature capability, and good adhesion for the coated substrate are used for selecting this type of painting.

7. COATING TECHNIQUES

  i. **Spraying** – This technique is the first choice for coating application because the coating covers the entire surface easily and without shading.
 ii. **Dip** – If the product needs to coat, it can dip easily into the coating liquid. This method gives good coverage around all the edges neatly.
iii. **Brush** – Brushing is the most common method for all types of interior and exterior materials. But this method is suitable for low volume and careful coating type on the electronic board.

CONCLUSION

The corrosion arises in the electronic devices and printed circuit board due to some environmental conditions like humidity, moisture and some other chemical reactions of acid and base. The acidic nature can help the circuit board gets an oxidation process or corrosion. The corrosion types are discussed briefly. In addition, we noticed some protective mechanisms for reducing corrosion.

CONSENT FOR PUBLICATION

Not applicable.

CONFLICT OF INTEREST

The author declares no conflict of interest, financial or otherwise.

ACKNOWLEDGEMENTS

Declared none.

REFERENCES

[1]   Vimala, J.S.; Natesan, M.; Rajendran, S. Corrosion and protection of electronic components in different environmental conditions an overview. *The Open Corrosion Journal,* **2009**, *2*, 105-11.

[2]   Ambat, R. *A review of Corrosion and environmental effects on electronics*; Manufacturing and Management, **2009**.

[3]   Culen, D. *Preventing Corrosion of PCB Assemblies On Board Technology,* **2008**.

[4]   Schueller, R. *Considerations for selecting a printed circuit board surface finish DfR Solutions,*

[5]   Kanegsberg, B.; Kanegsberg, E. *Handbook for Critical Cleaning. Applications, processes, and controls,* 2nd ed; CRC Press, **2011**, p. 524.

[6]   Seeling, K.; O'Neill, T. Conformal Coating over No-Clean Flux. *SMT Surf. Mt. Technol. Mag.,* **2014**, *29*, 52-58.

[7]   Tencer, M.; Moss, J.S. Humidity Management of Outdoor Electronic Equipment: Methods, Pitfalls, and Recommendations. *IEEE Trans. Compon. Packag. Tech.,* **2002**, *25*, 66-72. [http://dx.doi.org/10.1109/6144.991177]

[8]   Jacobsen, J.B.; Krog, J.P.; Holm, A.H.; Rimestad, L.; Riis, A. Climate-Protective Packaging: Using basic physics to solve climatic challenges for electronics in demanding applications. *IEEE Ind. Electron. Mag.,* **2014**, *8*(3), 51-59. [http://dx.doi.org/10.1109/MIE.2014.2330912]

[9]   Mauer, L.J.; Taylor, L.S. Water-solids interactions: deliquescence. *Annu. Rev. Food Sci. Technol.,* **2010**, *1*, 41-63. [http://dx.doi.org/10.1146/annurev.food.080708.100915] [PMID: 22129329]

[10]   Fan, X.J.; Lee, S.W.R.; Han, Q. Experimental investigations and model study of moisture behaviors in polymeric materials. *Microelectron. Reliab.,* **2009**, *49*(8), 861-871. [http://dx.doi.org/10.1016/j.microrel.2009.03.006]

[11]   Wong, E.H.; Koh, S.W.; Lee, K.H.; Rajoo, R. Comprehensive treatment of moisture induced failure - Recent advances. *IEEE Trans. Electron. Packag. Manuf.,* **2002**, *25*, 223-230. [http://dx.doi.org/10.1109/TEPM.2002.804613]

[12]   Frémont, H.; Horaud, W.; Weide-Zaage, K. Measurements and FE-simulations of moisture distribution in FR4 based printed circuit boards *7th International Conference on Thermal, Mechanical and Multi physics Simulation and Experiments in Micro-Electronics and Micro-Systems,* **2006**, pp. 1-6. [http://dx.doi.org/10.1109/ESIME.2006.1643964]

[13]   Xuejun, F.; Jiang, Z.; Chandra, A. Package structural integrity analysis considering moisture. *Proc of Electronic Components and Technology Conference,* (58th ECTC.) **2008**, pp. 1054-1066. [http://dx.doi.org/10.1109/ECTC.2008.4550106]

[14]   Lai, M.; Botsis, J.; Cugnoni, J.; Coric, D. An experimental-numerical study of moisture absorption in an epoxy. *Compos., Part A Appl. Sci. Manuf.,* **2012**, *43*(7), 1053-1060. [http://dx.doi.org/10.1016/j.compositesa.2012.01.027]

[15]   Smith, B.A.; Turbini, L.J. Characterizing the weak organic acids used in low solids fluxes. *J. Electron. Mater.,* **1999**, *28*, 1299-1306. [http://dx.doi.org/10.1007/s11664-999-0171-2]

[16]   Qu, G.; Vegunta, S.S.S.; Mai, K.; Weinman, C.J.; Ghosh, T.; Wu, W.; Flake, J.C. Copper oxide removal activity in non-aqueous carboxylic acid solutions. *J. Electrochemist.,* **2013**, *160*, E49-E530. [http://dx.doi.org/10.1149/2.083304jes]

[17]   Minzari, D.; Jellesen, M.S.; Møller, P.; Ambat, R. Morphological study of silver corrosion in highly aggressive sulfur environments. *Eng. Fail. Anal.,* **2011**, *18*(8), 2126-2136. [http://dx.doi.org/10.1016/j.engfailanal.2011.07.003]

[18]   Tegehall, P-E. *Impact of Humidity and Contamination on Surface Insulation Resistance and Electrochemical Migration in The ELFNET Book on Failure Mechanisms, Testing Methods, and Quality Issues of Lead-Free Solder Interconnects*; Springer, **2011**, pp. 227-253.

[19]   Perrone, M.G.; Larsen, B.R.; Ferrero, L.; Sangiorgi, G.; De Gennaro, G.; Udisti, R.; Zangrando, R. a. Gambaro, and E. Bolzacchini, Sources of high PM2.5 concentrations in Milan, Northern Italy: Molecular marker data and CMB modeling. *Sci. Total Environ.,* **2012**, *414*, 343-355.
[http://dx.doi.org/10.1016/j.scitotenv.2011.11.026] [PMID: 22155277]

[20]   Sinclair, J.D.; Psota-Kelty, L.A.; Weschler, C.J.; Shields, H.C. Measurement and modeling of air bone concentrations and indoor surface accumulation rates of ionic substances at Neenah, Wisconsin. *Atmos. Environ., A Gen. Topics,* **1990**, *24*, 627-638.
[http://dx.doi.org/10.1016/0960-1686(90)90018-I]

[21]   Litvak, A.; Gadgil, A.J.; Fisk, W.J. Hygroscopic fine mode particle deposition on electronic circuits and resulting degradation of circuit performance: an experimental study. *Indoor Air,* **2000**, *10*(1), 47-56.
[http://dx.doi.org/10.1034/j.1600-0668.2000.010001047.x] [PMID: 10842460]

[22]   Tencer, M. Deposition of aerosol ('hygroscopic dust') on electronics - Mechanism and risk. *Microelectron. Reliab.,* **2008**, *48*, 584-593.
[http://dx.doi.org/10.1016/j.microrel.2007.10.003]

[23]   Lohbeck, D. Design for dust. *Test Meas. World,* **2008**, *28*, 47-52.

[24]   Gaseous and Particulate Contamination Guidelines for Data Centers Am. Soc. Heating, Refrig. *Air-Conditioning Eng.,* **2011**, *2011*, 1-22.

[25]   Song, B.; Azarian, M. H.; Pecht, M. G. Effect of temperature and relative humidity on the impedance degradation of dust-contaminated electronics. *J. Electrochem. Soc.,* **2013**, *160*(3), C97-C105.
[http://dx.doi.org/10.1149/2.024303jes]

[26]   Frankenthal, R.P.; Siconolfi, D.J.; Sinclair, J.D. Accelerated Life Testing of Electronic Devices by Atmospheric Particles: Why and How. *J. Electrochem. Soc.,* **1993**, *140*, 3129-3134.
[http://dx.doi.org/10.1149/1.2220997]

[27]   Cooper, D.W. Particulate Contamination and Microelectronics Manufacturing: An Introduction. *Aerosol Sci. Technol.,* **2015**, *5*, 287-299.
[http://dx.doi.org/10.1080/02786828608959094]

[28]   Thomas, D.A.; Avers, K.; Pecht, M. The 'trouble not identified' phenomenon in automotive electronics. *Microelectron. Reliab.,* **2002**, *42*, 641-651.
[http://dx.doi.org/10.1016/S0026-2714(02)00040-9]

[29]   Qi, H.; Ganesan, S.; Pecht, M. No-fault-found and intermittent failures in electronic products. *Microelectron. Reliab.,* **2008**, *48*, 663-674.
[http://dx.doi.org/10.1016/j.microrel.2008.02.003]

CHAPTER 4

Corrosion of Polymer Materials

U. Naresh[1], N. Suresh Kumar[2,*], K. Chandra Babu Naidu[3], B. Venkata Shiva Reddy[3, 4], A. Manohar[5] and T. Anil Babu[3]

[1] *Department of Physics, BIT Institute of Technology, Hindhupur, A.P., India*

[2] *Department of Physics, JNTU College of Engineering Anantapur, Anantapuramu-515002, A.P., India*

[3] *Department of Physics, GITAM Deemed to be University , Bangalore - 562163, Karnataka, India*

[4] *Department of Physics, The National College, Bagepalli-561207, Karnataka, India*

[5] *Department of Materials Science and Engineering, Korea University, 145 Anaro, Seongbukgu, Seoul 02841, Republic of Korea*

Abstract: The technology is growing very fast in the world with the expansion of novel functional materials. These novel materials are the backbone for technology development. Whereas the lifetime of materials is decreased due to environmental aspects such as the breakdown of a chemical bond with the oxidation process. Usually, it may be called corrosion. In this chapter, we reviewed some corrosion aspects like corrosion types and causes of corrosion. In addition, it also discusses protective materials like conductive polymer and its composites, such as conducting polymers mixers of magnetic materials and graphene additives.

Keywords: Conducting, Dealloying, Fretting, Hygroscopic Contaminants, Monomers, Nanoparticles, Polymers.

1. INTRODUCTION

It is a well-known fact that the corrosion is treated as the morphological imperfection of metal-based materials because of the oxidation process. Generally, polymers or some lubricants are used to protect metals from corrosion. The corrosion usually happens on the polymers, and rubber materials. This type of corrosion is very different than natural corrosion. Because, other corrosions can be identified easily, whereas the polymer corrosion is not possible to identify quickly. The probable chance to find corrosion in polymers is the loss of

* **Corresponding author N. Suresh Kumar:** Department of Physics, JNTU College of Engineering, Anantapur-515002, Andhra Pradesh, India; Tel: +91-81211 27157; E-mail: sureshmsc6@gmail.com

N. Suresh Kumar, P. Banerjee, H. Manjunatha and K. Chandra Babu Naidu (Eds.)

mechanical properties of the materials. That is, if the mechanical stress is applied to the corrosion polymer materials and their chemical environment, those materials immediately undergo creeping or cracks on the surfaces. Generally, chemical reaction and physical interactions are the two major types occurred in polymer materials.

2. CHEMICAL REACTION

The polymer materials are non-crystalline materials having the mixture of molecular series of carbon, oxygen, hydrogen. As the materials are exposed in the environment, the chemical reactions take place leading to the corrosion. If the reaction takes place under the heat or far infrared and UV-Visible radiation, the polymer chain gets breakdown. In the environment, the ozone and water make hydrolysis process of polymer monomers. Therefore, it leads to the breakdown of chemical bonds. The difference between breakdowns of chemical bond is due to inflammation or dissolving chemical elements without variation of structure.

Excess of reaction resistance is required on the molecular structure of the polymer, whether they are unstructured or partially crystalline. Semi-crystalline materials are more resistant than amorphous polymers. Physical contacts on polymers happen mainly due to the environment. Hence, the materials undergo dissolving or swelling because of chemical reactions or leakage of polymers. However, the interaction depends on the penetration of additives into polymer substances. More generally, organic polymers usually affect many physical interactions, whereas some of the materials like strong acids and base reactance may cause the breakdown of the polymer materials. Interestingly, all the conducting polymers acquired impressible importance in the field of coating technology, as they show very good corrosion resistance properties. Especially polyaniline coatings are the considerable choice for protecting metals. The blend type polyaniline additives are one of the best choices to protect metals from corrosion, as they provide many benefits [1 - 4]. The polyaniline emeraldine salt is the most thoroughly investigated and studied polymer.

3. CAUSES OF CORROSION

3.1. Moisture

The printed electronic board assembly surface gets corrosion mainly due to humidity interaction by creating a water layer. There are several ways to react humidity with PCBA. The humidity interactions with metals or solid-water interactions can be classified into surface interactions, condensed water interactions and water internalized into the solid. The humidity interactions can be classified as deliquescence, adsorptions, capillary condensations, absorption of water, and formation of the layer. Among all the types, deliquescence, adsorption, and capillary condensation are the important interactions of PCBA. Of course, the PCBA material, architecture, surface contaminations are also dependable parts for corrosion. Water penetration through the absorption modification is a part of properties of PCB laminated or bulk components and, therefore, equipment causes internal degradation [5].

3.2. Water Absorption

The water absorption is usually caused due to penetration of wetness in PCBA and further it forms the layers of dampness. The moisture ingress prior to soldering can cause delamination at the later stage of manufacturing. An example of this is reflow soldering of the surface mount components (SMCs). In addition, as the electronic device is warmed up, water absorption affects the thermal properties of PCB materials. The dielectric property of laminate is increased due to interface leakage. Hence, the efficiency in multi-layer PCBA is also increased [6]. Generally, the capacitance of the capacitor changes the electronic board laminates and the moisture content in the laminates follows a proportionality relationship for the moisture measurement [7 - 9]. At the time of assembly process, the circuitry is confined to interacting the solder by applying a solder mask, thereby covering most of the laminate surface.

3.3. Water Adsorption by the Hygroscopic Contaminants

Absorption by water-soluble ionic residues and contaminants on the surface of the PCBA affects the surface insulation resistance (SIR). The number of adsorbed molecular layers of water increases with the relative humidity. As the ambient relative humidity meets a cutoff value of relative humidity, the deliquescence (conversion of solid into saturated solution it is generally referred as phase transformation and it depends upon the atmosphere characteristics, that is temperature) will be formed in the PCBA. At the stage of relative humidity of deliquescence, the moisture may be transformed as temperature dependent solid.

4. TYPES OF CORROSION

4.1. General Corrosion

This is also called uniform corrosion. It is well recognized corrosion, and the developed synthetic and chemical reactions are due to the fragile of the uncovered surface. Hence, the metal gets apart of degradation. General corrosion corresponds to the calculation of metal deprivation by corrosion. However, it is considered a safety corrosion.

4.2. Galvanic Corrosion

This corrosion is also called bimetallic corrosion. As the two metallic elements soldered with each other and used as a destructive electrolyte, those bimetals get oxidation, and it loses its nature. Then, galvanic corrosion is formed. In these metal electrolytes, one is used as the cathode (disintegrate more gradually) and another one is used as anode (chances get corrosion faster). Hence, precautionary decisions must be taken including the good selection of the material at the time of fabrication. General reasons for galvanic corrosion are considered as choosing reactive electrochemical material, the electrical contact and the metals used as the electrolyte.

4.3. Pitting Corrosion

The pitting type of corrosion is a phenomenon of creating small regain of a hole or dot type corrosion on the surface of the metal. This corrosion is generally considered as part of galvanic corrosion as it starts from cathode and ends at the catholic point. The degeneration process starts with a prompt failure of small region. Identifying the pitting corrosion is very difficult because it starts in a very small area and spreads quickly. Even it can spread in the case of the material coated with any kind of polymer coating. Hence, it may consider as polymer corrosion. This corrosion can be protected using anodic and cathodic prevention by using advanced alloys.

4.4. Dealloying

Metallic alloys or composites are usually exposed to corrosive familiarity. Then, one of the elements in alloy gets breakage of bonding of alloys. For example, in the alloy of brass, copper and zinc, dealloying or dezincification normally arises in a corrosive medium. Then, the color of brass is converted to reddish from yellow. We can also find the leaving iron from graphite network. Precious metal alloys such as gold containing copper or silver, sulfide environment, the preferential dissolution of copper or silver will be taken place.

4.5. Erosion Corrosion

The disintegration of metals or alloy is happened due to outer surfaces and corrosive fluids kind of environment for the coated materials. The scratches type of damage is started at the points where erosion takes place. Then, the rate of corrosion is rapidly increased. For example, water tube walls of stainless-steel alloys will be exposed to moving fluids to make them are flat due to erosion-corrosion.

4.6. Fretting

If the high weight or load materials are used (vibrated) regularly, they become the main sources of the fretting type of corrosion. This corrosion initially begins with the formation of pits, grooves, and polymer coated materials. For the protection of these materials, lubricants are necessary. Hardening of the surface can also become advantageous. Seals should be used to withstand vibrations and they keep it away from moisture and oxygen. Fig. (**1**) shows fretting corrosion.

Fig. (1). Fretting corrosion.

5. CORROSION PROTECTION-CONDUCTING POLYMERS

Ferric composites like mild steel are one of the repeated sources for the corrosion, since it is used for long days in the machinery factories as constructing substance because of flexibility and stability without cracking. These types of materials will be bent and collapse at the time of accidents like earthquakes or some unexpected disasters. The most capable way to protect mild steel or its complexes is the corrosion resistance coatings with sufficient polymers. That too, by using the conducting polymer composites, the materials will be easily converted as corrosion-resistant materials. Nowadays, researchers focused to improve the coating effort for many commercial applications. From this point of view, we reviewed coatings which are made with the polymers and composites [10 - 12].

Different applications in the technology caused for the investigation of conducting polymers effectively in these days. Especially in the field of energy storage system, memory storage devices, catalysis, drug delivery, sensing technology and anticorrosion coating technology. The coating rate depends on the corrosion rate or efficiency of corrosion of the material within the corrosive environment. In some cases, the physical properties such as electrical, structural properties may prejudice with the effect of polymer coating level [13, 14]. In the group of conducting polymers, some of the polymers like polyaniline and poly-pyrrole are the main polymers for the various applications. The anticorrosion properties of conducting polymers increased the importance of CPs with their own interesting properties [15]. In the group of CPs, polyaniline is the most studied polymer due to its specialized properties, its nature control of electrical conduction and stability in a chemical reaction with the environment [16 - 18]. Polyaniline exists in the form of oxidation such as salt, base, leucoemraldin and pernigraniline. Spinks *et al.* [19, 20] verified the ability of both PANI and PPy coatings to test the corrosion resistance of steel. The experimental evidence demonstrates that the PC galvanic coupled to the metal is a reason for an anodic shift in the corrosion. Moreover, some authors argue that this shift causes passivation and anodic protection. This mechanism is supported by studies showing stable oxide formation on the metal surface after contact with the conducting polymer. Polyaniline and its composites like polyaniline and magnetic nanoparticle, ceramic materials and carbon derivatives are discussed in this chapter.

5.1. Poly-aniline- Magnetic Nanoparticles Coatings

Alves *et al.* [21] studied the polyaniline coating for painting on mild steel protection for corrosion protection. The prepared samples were characterized with the help of FTIR and Raman spectroscopy. In this study, no degradation is found in the polymer when it is integrated into the paint. The synthesized samples show electro-activity in 1 mole of HCl, and some of the samples are shown in NaCl with a limited percentage. The results show clearly that doped polyaniline impr-oves the corrosion resistance capacity. The apparatus Raman spectroscopy resulted from the sample of magnetic particle and polyaniline composites establishing an oxide layer in between interface of coating. This indicates that the composites do not set corrosion chances. The polyaniline magnetic particle coatings showed that the presence oxide layer is formed by the oxide. The Fe_2O_3 based materials showed less performance in the corrosion [22]. In their investigation, the prepared samples show impressive corrosion resistance performance for metals. The boundary conditions on the metal surfaces and better versatility in coatings make many applications despite temperature conditions.

The most studied magnetic ferrites are NiZn ferrite due to their better magneto-resistive nature and dielectric properties. In the case of conducting polymers, polyaniline is attracted much attention because of its chemical, and physical properties similar like metals. Both materials in nature are chemically stable and soluble in various known solvents. The behavior of polyaniline shows the high density exhibiting the magnetic nature. This magnetic nature of polyaniline reflects various applications like waveguide generators, transducers, sensors, wave interference filters, various magnetic components, *etc.*

This type of material becomes the class of mixture coatings for corrosion resistive application. The NiZn ferrite and PANI mixture can show anti-corrosive properties. These are evaluated by salt-spray test and electromechanical test. The results reveal good corrosive resistive properties. In the electromechanical studies, the low current density expressed that the barrier is formed between substrate surface and water. The NMR results and potential dynamic polarization curve stated that the prepared hybrid coatings became denser than previous, as NiZn ferrites and PANI particles mixture incorporated on the metallic surface. The nickel ferrite is the most efficient material for high-frequency applications due to its versatile properties. Not only nickel ferrite but also some permanent magnets like gadolinium ferrite contains very good applications in different fields. In previous work, gadolinium ferrite and conducting polymer combination were prepared through *"in-situ"* polymerization technique. The synthesized material is tested for the anti-corrosive properties and results revealed good anti-corrosive properties. In these aspects, the materials of conducting polymers (polyaniline, polypyrrole) and magnetic materials (Fe_2O_3, Fe_2O_4) are used for anti-corrosive properties [23, 24].

The magnetic properties like coercivity, retentivity, *etc.*, obtained from hysteresis curves, show a vital role in coatings. The coercivity of the magnetic particles found in the PANI and ferrite showed the ferrimagnetic nature. It can demonstrate that the non-magnetic nature of PANI or the interaction of magnetic particles in the ferrite composites. The electrical studies of these samples revealed that the variation of conductivity in ferrite content shows a smaller value. The existence of coating application is observed by polymerization of PANI. When using the anti-coating application, the bulk nanoclusters can be made with nanoparticles and metal oxides [25]. The deposition of L-b-L is one of the most promising methods for the synthesis of intelligent anti-corrosion nano-retainers [26 - 29]. In the L-b-L method, the core of nano-retina is essentially solid prepared from metal oxides and/or inorganic/organic compounds. The solid cores in the L-B-L nano-storage contain much compensation, such as maintaining the boundary properties of the coating after electrolyte scattering, reducing reservoir losses during the dispersion stage of the natural coating, and make easy for the synthesis of reservoirs in the nanoscale range. Using L-b-L strategy, various layers can be applied to the solid core. Further, the corrosion inhibitors can be loaded between the layers. Finally, it will cover the polymer coating layers and prevent the release of the resistor from the nano stores before use. The shell of the nanoclusters can be prepared with the assist of an *in-situ* emulsion polymerization [30]. Polyelectrolytes are a class of polymer materials and further divided into two kinds together with polycations and polyanions. In the electrolytes, the polyanions consist of a greater amount of ionizable groups of acids. These are generally ionized with alkaline and neutral solutions [31]. There are some layers consisting of polyelectrolytes, polycations and polyanion layers used for the coating applications. Due to this attractive force in between the layers and molecules, the polyelectrolyte layer is not charged. The pH value is also a reason for the corrosion resistance behavior as forming a polycation layer. The core mechanism of inhibiting corrosion is well established one, wherein the inhibitors are being created in the nucleus due to the strong attraction between opposition-loaded poly-cations and corrosion inhibitors. The polyaniline (PANI) is a conjugated polymer having economic & environmental stability and double bond with a conductive nature [32 - 34]. The polyaniline is not charged in some alkaline, neutral solutions and corrosion resistance polymers released from the ferrite structure. The use of PANI as a corrosion inhibitor is reported in many kinds of literature [35 - 38].

5.2. Poly-aniline-Carbon Based Materials

Graphene is a thin layer of carbon derivative having honeycomb structure with sp3 and sp^2 configuration. This type of unique structure enables the generation of various applications in technology [39]. Whereas one of the graphene derivatives is graphene oxide, and it can be dispersed in different solvents for applications. Organic corrosive resistance coating prepared for the metallic substrate surfaces has become important from the last few decades to enable technology development. The literature suggested that the coating materials are prepared through various materials like TiO_2, SiO_2, ZnO_2 and ferrites, magnetic materials, *etc.* [40, 41]. At the same time, some other materials that are mixers of carbon-based materials like carbon nanotubes, graphene composites of PANI are tested for corrosion resistance materials [42, 43]. The graphene is carbon derived material having potential applications in many different areas. There are several researchers studying the properties of graphene and its derivatives. The researchers found different synthesis techniques and application for the anticorrosive nature [44]. The most recent studies reported that the graphene is considered as an applicable material for the control of corrosion of polymers mixers. The unique properties like chemical inertness, flexibility, chemical stability, and thermal stability of the graphene suggested that the graphene materials are suitable for anti-corrosion application. In this direction, Chang *et al.* [45] demonstrated that the graphene and polymer composites enable the increase of anti-corrosive nature for the steel materials. Yu *et al.* [46], synthesized the graphene oxide and polystyrene composites by *in-situ* polymerization technique for superior corrosion protection. Singh *et al.* [47], provided the result about graphene layered coating and observed its superior anti-corrosive nature. The cathodic electrophoretic deposition of graphene layer coating got more advantages than usual chemical vapor deposition as its advances like controlled nanostructure on a surface of large range substrates.

The study about the Pani-aniline (PANI), the anti-corrosive nature of Cu and graphene composite coating is done by Jafari *et al.* [48]. The cyclic voltammetry (CV) method is used for the formation of PANI and graphene mixer on the surface of the Cu substrate. The coating efficiency is checked by the EIS and polymerization technique. The investigations suggested the increment of corrosive resistance nature against Cu corrosion. The morphology of prepared samples is checked through the SEM and EIS technique. The morphology obtained by SEM and EIS revealed that polyaniline is distributed uniformly with a decrease of polymer pores. The study evidenced that the coated Cu substrate contains graphene polyaniline nanostructures uniformly on the surface. The polymer pores are decreased as incorporating graphene on the Cu substrate. The corrosion study is carried out as the samples are tested in acidic and base solutions. The graphene and polyaniline mixers could not show the damages of Cu surface. PANI/G nanocomposite coatings show outstanding anti-corrosion in an aggressive environment. Therefore, the use of nanoparticles of polyaniline and graphene in a composite material increased the surface anthropolis performance significantly. Results showed that the use of polyaniline with graphene nanoparticles contributes to the formation of a composite layer, thereby reducing the corrosion efficiency of the metal surface to lower values and reducing the corrosion rate.

The atmospheric parameters are much important for the corrosive environment. The related equipment pertains to the most effective parameters for corrosion leading to failure of equipment measurements. This correlation between weathering and corrosion will not end, and this leads to the development of new materials for the anticorrosive nature. The graphene-based materials become a good choice for anti-corrosive materials. Thus, from the last two decades, graphene-related materials research has been growing well.

Amrollahi *et al.* [49] polymerized polyaniline on the surface of graphene oxide in the *in-situ* process. These synthesized graphene oxide polyaniline sheets are tested with FTIR, XRD, Raman, UV–Visible and FE-SEM for structural and morphological properties. The DPPPH experimental results revealed corrosive capabilities of graphene oxide due to the presence of polyaniline. The weathering and corrosive properties are also increased with graphene oxide and polyaniline coatings. The weathered and corroded epoxy coatings are evaluated by different approaches. All the results depicted a wide range of protective properties against corrosive in different weathering conditions by introducing GO-PANI composites [50]. The positive brunt of polyaniline on the graphene oxide surface is also a shielding for the UV-radiation, radical scavenging, and barrier properties. It is the main mechanism involved in the simultaneous promotion of weathering and corrosion performance [51].

CONCLUSION

In this chapter, we discussed the kinds of corrosions and protective materials. In addition, we focused on the new materials which are drawn by the various researchers for corrosion prevention. The conducting polymers and their composites like polyaniline polypyrrole ferrite mixers and graphene-derived combinations are discussed briefly.

CONSENT FOR PUBLICATION

Not applicable.

CONFLICT OF INTEREST

The author declares no conflict of interest, financial or otherwise.

ACKNOWLEDGEMENTS

Declared none.

REFERENCES

[1]   Meng, F.D.; Liu, L.; Tian, W.L.; Wu, H.; Li, Y.; Zhang, T.; Wang, F.H. The influence of the chemically bonded interface between fillers and binder on the failure behaviour of an epoxy coating under marine alternating hydrostatic pressure. *Corros. Sci.,* **2015**, *101*, 139-154.
      [http://dx.doi.org/10.1016/j.corsci.2015.09.011]

[2]   Andreeva, D.V.; Fix, D.; Möhwald, H.; Shchukin, D.G. Buffering polyelectrolyte multi- layers for active corrosion protection. *J. Mater. Chem.,* **2008**, 1738-1740.

[http://dx.doi.org/10.1039/b801314d]

[3]   Ghelichkhah, Z.; Sharifi-Asl, S.; Farhadi, K.; Banisaied, S.; Ahmadi, S.; Macdonald, D.D. L-cysteine polydopamine nanoparticle-coatings for copper corrosion protection. *Corros. Sci.,* **2015**, 129-139.
[http://dx.doi.org/10.1016/j.corsci.2014.11.011]

[4]   Sababi, M.; Pan, J.; Augustsson, P-E.; Sundell, P-E.; Claesson, P.M. Influence of polyaniline and ceria nano particle additives on corrosion protection of a UV cure coating on carbon st. *Corros. Sci.,* **2014**, *84*, 189-197.
[http://dx.doi.org/10.1016/j.corsci.2014.03.031]

[5]   Sepideh, P.; Farhad, S.; Jizhou, D.; Alimorad, R.; Fang, G.; ElhamGarmroudi, N.; Baorong, H. Polymer/Inorganic nano-composite coatings with superior corrosion protection performance: A review. *J. Ind. Eng. Chem.,* **2020**, *88*, 29-57.
[http://dx.doi.org/10.1016/j.jiec.2020.04.029]

[6]   Anuradha, K.; Vimala, R.; Narayanasamy, B.; Arockiaselvi, J.; Rajendran, S. Corrosion inhibition of carbon steel in low chloride media by an aqueous extract of Hibiscus rosa-sinensislinn. *Chem. Eng. Commun.,* **2008**, *195*, 352-366.
[http://dx.doi.org/10.1080/00986440701673283]

[7]   Guerreiro da Trindade, L.R.; Simoes, G. Evidence of caffeine adsorption on a lowcarbon steel surface in ethanol. *Corros. Sci.,* **2009**, *51*, 1578-1583.
[http://dx.doi.org/10.1016/j.corsci.2009.03.038]

[8]   Li, Y.; Zhao, P.; Liang, Q.; Hou, B. Berberine as a natural source inhibitor for mild steel in 1M H2SO4. *Appl. Surf. Sci.,* **2005**, *252*, 1245-1253.
[http://dx.doi.org/10.1016/j.apsusc.2005.02.094]

[9]   De Souza, F.S.; Spinelli, A. Caffeic acid as a green corrosion inhibitor for mild steel. *Corros. Sci.,* **2009**, *51*, 642-649.
[http://dx.doi.org/10.1016/j.corsci.2008.12.013]

[10]   Izadi, M.; Shahrabi, T.; Ramezanzadeh, B. Synthesis and characterization of an advanced layer-b- -layer assembled Fe_3O_4/polyanilinenano reservoir filled with Nettle extract as a green corrosion protective system. *J. Ind. Eng. Chem.,* **2018**, *57*, 263-274.
[http://dx.doi.org/10.1016/j.jiec.2017.08.032]

[11]   Synthesis of polyaniline-modified graphene oxide for obtaining a high performance epoxy nano-composite film with excellent UV blocking/ anti-oxidant/ anti-corrosion capabilities. *Composites Part B*, **2019**, *173*, 106804.

[12]   De Souza, S. Smart coating based on polyaniline acrylic blend for corrosion protection of different metals. *Surf. Coat. Tech.,* **2007**, *201*, 7574-7581.
[http://dx.doi.org/10.1016/j.surfcoat.2007.02.027]

[13]   Stejskal, J.I. Sapurina, M. Trchová, E.N. Konyushenko, Oxidation of aniline: polyaniline granules, nanotubes, and oligoaniline microspheres. *Macromolecules,* **2008**, *41*, 3530-3536.
[http://dx.doi.org/10.1021/ma702601q]

[14]   Grgur, B.N.; Elkais, A.R.; Gvozdenovi, M.M.; Drmani, S.Z.; Lj, T. Tri῀sovi, B.Z. Jugovi Corrosion of mild steel with composite polyaniline coatings using different formulations. *Prog. Org. Coat.,* **2015**, *79*, 17-24.
[http://dx.doi.org/10.1016/j.porgcoat.2014.10.013]

[15]   Kamburovaa, K.; Boshkovaa, N.; Tabakovab, N.; Boshkova, N.; Radeva, Ts. Corrosion. Polymer/Inorganic nano composite coatings with superior corrosion protection performance: A review. *J. Ind. Eng. Chem.,* **2020**, *88*, 29-57.
[http://dx.doi.org/10.1016/j.jiec.2020.04.029]

[16]   McCafferty, E. *Introduction to Corrosion Science*; Springer: Alexandria, VA, **2010**, p. 199.
[http://dx.doi.org/10.1007/978-1-4419-0455-3]

[17]   De Albuquerque, J.E.; Mattoso, L.H.C.; Faria, R.M.; Masters, J.G.; Mac Diarmid, A.G. Study of the interconversion of polyaniline oxidation states by optical absorption spectroscopy. *Synth. Met.,* **2004,** *146,* 1-10.
[http://dx.doi.org/10.1016/j.synthmet.2004.05.019]

[18]   Dupare, D.B.; Shirsat, M.D.; Aswar, A.S. Inorganic acids doped PANI-PVA composites films as a gas sensor. *Pac. J. Sci. Technol.,* **2009,** *10*(1), 417-422.

[19]   Jin, E.; Liu, N.; Lu, X.; Zhang, W. Novel micro/nanostructures of polyaniline in the presence of different amino acids *via* a self-assembly process. *Chem. Lett.,* **2007,** *36,* 1288-1289.
[http://dx.doi.org/10.1246/cl.2007.1288]

[20]   Lia, Y.; Wang, B.; Feng, W. Chiral polyaniline with flaky, spherical and urchin-like morphologies synthesized in the l-phenylalanine saturated solutions. *Synth. Met.,* **2009,** *159*(15), 1597-1602.
[http://dx.doi.org/10.1016/j.synthmet.2009.04.023]

[21]   Li, X.; Wang, X.; Zhang, L.; Lee, S.; Dai, H. Chemically derived, ultrasmooth graphene nanoribbon semiconductors. *Science,* **2008,** *319*(5867), 1229-1232.
[http://dx.doi.org/10.1126/science.1150878] [PMID: 18218865]

[22]   Shen, J.; Hu, Y.; Shi, M.; Lu, X.; Qin, C.; Li, C. Fast and facile preparation of graphene oxide and reduced graphene oxide nano platelets. *Chem. Mater.,* **2009,** *21,* 3514-3520.
[http://dx.doi.org/10.1021/cm901247t]

[23]   Mahsa, D.; Ebrahim, G.; Bahram, R.; Mohammad, M. Designing a zinc-encapsulated Feldspar as a unique rock-forming tecto-silicate nano container in the epoxy coating; improving the robust barrier and self-healing anti-corrosion properties. *Constr. Build. Mater.,* **2020,** *243,* 118-215.

[24]   Wu, K.H.; Chao, C.M.; Liu, C.H.; Chang, T.C. Characterization and corrosion resistance of organically modified silicate–NiZn ferrite/ polyaniline hybrid coatings on aluminum alloys. *Corros. Sci.,* **2007,** *49,* 3001-3014.
[http://dx.doi.org/10.1016/j.corsci.2007.02.008]

[25]   Aphesteguy, J.C.; Bercoff, P.G.; Jacobo, S.E. Preparation of magnetic and conductive Ni–Gd ferrite-polyaniline composite. *Physica B,* **2007,** *398,* 200-203.
[http://dx.doi.org/10.1016/j.physb.2007.04.018]

[26]   Grigoriev, D.O.; Koehler, K.; Skorb, E.; Shchukin, D.G.; Moehwald, H. Polyelectrolyte complex as a smart depot for self-healing anticorrosion coatings. *Soft Matter,* **2009,** *5,* 1426-1432.
[http://dx.doi.org/10.1039/b815147d]

[27]   Zheludkevich, M.L.; Shchukin, D.G.; Yasakau, K.A.; Moehwald, H.; Ferreira, M.G.S. Anticorrosion coatings with self-healing effect based on nano containers impregnated with corrosion inhibitor. *Chem. Mater.,* **2007,** *19,* 402-411.
[http://dx.doi.org/10.1021/cm062066k]

[28]   Bhanvase, B.A.; Kutbuddin, Y.; Borse, R.N.; Selokar, N.R.; Pinjari, D.V.; Gogate, P.R.; Sonawane, S.H. Ultrasound assisted synthesis of calcium zinc phosphate pigment and its application in nano container for active anticorrosion coatings. *Chem. Eng. J.,* **2013,** *231,* 345-354.
[http://dx.doi.org/10.1016/j.cej.2013.07.030]

[29]   Falcon, J.M.; Batista, F.F.; Aoki, I.V. Encapsulation of dodecylamine corrosion inhibitor on silica nanoparticles. *ElectrochemicaActa,* **2013,** *124,* 109-118.
[http://dx.doi.org/10.1016/j.electacta.2013.06.114]

[30]   Bhanvase, B.A.; Patel, M.A.; Sonawane, S.H. Kinetic properties of layer-by-layer assembled cerium zinc molybdatenanocontainers during corrosion inhibition. *Corros. Sci.,* **2014,** *88,* 170-177.
[http://dx.doi.org/10.1016/j.corsci.2014.07.022]

[31]   Sonawane, S.H.; Bhanvase, B.A.; Jamali, A.A.; Dubey, S.K.; Kale, S.S.; Pinjari, D.V.; Kulkarni, R.D.; Gogate, P.R.; Pandit, A.B. Improved active anticorrosion coatings using layer-by-layer assembled ZnOnano containers with benzotriazole. *Chem. Eng. J.,* **2012,** *189-190,* 464-472.

[http://dx.doi.org/10.1016/j.cej.2012.02.076]

[32] Tyagi, M.; Bhanvase, B.A.; Pandharipande, S.L. Computational studies on release of corrosion inhibitor from layer-by-layer assembled silica nanocontainer. *Ind. Eng. Chem. Res.,* **2014**, *53*, 9764-9771.
[http://dx.doi.org/10.1021/ie5010064]

[33] Ghosh, S.K. *Functional Coatings*; WILEY-VCH: Weinheim, **2006**, p. 24.
[http://dx.doi.org/10.1002/3527608478]

[34] Wei, H.; Wang, Y.; Guo, J.; Shen, N.Z.; Jiang, D.; Zhang, X.; Yan, X.; Zhu, J.; Wang, Q.; Shao, L.; Lin, H.; Wei, S.; Guo, Z. Advanced micro/nanocapsules for self-healing smart anticorrosion coatings. *J. Mater. Chem. A Mater. Energy Sustain.,* **2014**, *3*, 469-480.
[http://dx.doi.org/10.1039/C4TA04791E]

[35] Chenan, A.; Ramya, S.; George, R.P.; Mudali, U.K. Hollow meso porous zirconia nano containers for storing and controlled releasing of corrosion inhibitors. *Ceram. Int.,* **2014**, *40*, 10457-10463.
[http://dx.doi.org/10.1016/j.ceramint.2014.03.016]

[36] Matthew, M. Ali, Julia C. Magee 1, Peter Y. Hsieh Corrosion protection of steel pipelines with metal-polymer composite barrier liners. *J. Nat. Gas Sci. Eng.,* **2020**, *81*, 103-407.

[37] Askari, M.; Aliofkhazraei, M.; Ghaffari, S.; Hajizadeh, A. Film former corrosion inhibitors for oil and gas pipelines - a technical review. *J. Nat. Gas Sci. Eng.,* **2018**, *58*, 92-114.
[http://dx.doi.org/10.1016/j.jngse.2018.07.025]

[38] Mokhatab, S.; Poe, W.A.; Mak, J.Y. *Handbook of Natural Gas Transmission and Processing: Principles and Practices*, 4th ed; Elsevier Science, **2018**.

[39] Bereket, G.; Hur, E.; Sahin, Y. Electro-deposition of polyaniline, poly (2-iodoaniline), and poly(aniline-co-2-iodoaniline) on steel surfaces and corrosion protection of steel. *Appl. Surf. Sci.,* **2005**, *252*, 1233-1244.
[http://dx.doi.org/10.1016/j.apsusc.2005.02.087]

[40] Izadi, M.T. Shahrabi, B. Ramezanzadeh, Synthesis and characterization of an advanced layer-by-layer assembled Fe_3O_4/polyanilinenano reservoir filled with Nettle extract as a green corrosion protective system. *J. Ind. Eng. Chem.,* **2018**, *57*, 263-274.
[http://dx.doi.org/10.1016/j.jiec.2017.08.032]

[41] Wrssling, B. Passivation of metals by coating with polyaniline: corrosion potential shift and morphological changes. *Adv. Mater.,* **1994**, *6*, 226-228.
[http://dx.doi.org/10.1002/adma.19940060309]

[42] Jeyaprabha, C.; Sathiyanarayanan, S.; Venkatachari, G. Co-adsorption effect of polyaniline and halide ions on the corrosion of iron in 0.5 M H_2SO_4 solutions. *J. Electroanal. Chem. (Lausanne Switz.),* **2005**, *583*, 232-240.
[http://dx.doi.org/10.1016/j.jelechem.2005.06.006]

[43] Fahlman, M.; Jasty, S.; Epstein, A.J. The role of Na-montmorillonite/cobalt ferrite nanoparticles in the corrosion of epoxy coated AA 3105 aluminum alloy. *Synth. Met.,* **2019**, *15*, 89-99.

[44] Kralji, M.; Mandi, Z.; Dui, L. Inhibition of steel corrosion by polyaniline coatings. *Corros. Sci.,* **2003**, *45*, 181-198.
[http://dx.doi.org/10.1016/S0010-938X(02)00083-5]

[45] Jamróz, E. Kulawik, P. Kopel, P. The Effect of Nanofillers on the Functional Properties of Biopolymer-Based Films: A Review. *Polymers (Basel),* **2019**, *11*, 675.
[http://dx.doi.org/10.3390/polym11040675]

[46] Roy, D.; Simon, G.P.; Forsyth, M.; Mardel, J. Towards a better understanding of the cathodic disbandment performance of polyethylene coatings on steel. *Adv. Polym. Technol.,* **2002**, *21*, 44-58.
[http://dx.doi.org/10.1002/adv.10010]

[47]   Chang, C.H.; Huang, T.C.; Peng, C.W.; Yeh, T.C.; Lu, H.I.; Hung, W.I.; Weng, C.J.; Yang, T.I.; Yeh, J.M. Novel anticorrosion coatings prepared from polyaniline/ graphene composites. *Carbon,* **2012**, *50*, 5044-5051.
[http://dx.doi.org/10.1016/j.carbon.2012.06.043]

[48]   Yu, Y.H.; Lin, Y.Y.; Lin, C.H.; Chan, C.C.; Huang, Y.C. High-performance polystyrene/ graphene-based nano-composites with excellent anti-corrosion properties. *Polym. Chem.,* **2014**, *5*, 535-550.
[http://dx.doi.org/10.1039/C3PY00825H]

[49]   Singh, P.; Nayak, S.; Nanda, K.K.; Jena, B.K.; Bhattacharjee, S.; Besra, L. The production of a corrosion resistant graphene reinforced composite coating on copper by electrophoresis deposition. *Carbon,* **2013**, *6*, 47-56.
[http://dx.doi.org/10.1016/j.carbon.2013.04.063]

[50]   Jafari, Y.S.M.; Shabani, N. Polyaniline/Graphenenano composite coatings on copper: Electro-polymerization, characterization, and evaluation of corrosion protection performance. *Synth. Met.,* **2016**, *217*, 220-230.
[http://dx.doi.org/10.1016/j.synthmet.2016.04.001]

[51]   Amrollahi, S.B.; Ramezanzadeh, H.; Yari, M. Synthesis of polyaniline modified graphene oxide for obtaining a high-performance epoxy nano-composite film with excellent UV blocking/ anti-oxidant/ anti-corrosion capabilities. *Compos., Part B Eng.,* **2019**, *173*, 106-804.
[http://dx.doi.org/10.1016/j.compositesb.2019.05.015]

CHAPTER 5

Corrosion and Corrosion Protection in Drinking Water Systems

T. Vidya Sagar[1], N. Suresh Kumar[2,*], K. Chandra Babu Naidu[3], B. Venkata Shiva Reddy[3, 4], D. Baba Basha[5] and T. Anil Babu[3]

[1] *Department of Physics, S K University, Anantapuramu-515003, A.P., India*

[2] *Department of Physics, JNTU College of Engineering Anantapur, Anantapuramu-515002, A.P., India*

[3] *Department of Physics, GITAM Deemed to be University, Bangalore - 562163, Karnataka, India*

[4] *Department of Physics, The National College, Bagepalli-561207, Karnataka, India*

[5] *Department of Physics, College of Computer and Information Sciences, Majmaah University, Al'Majmaah, Saudi Arabia*

Abstract: In this chapter, we discussed the corrosion and corrosion protection in drinking water systems. The corrosion of the drinking water distribution system (DWDS) basically depends on the pH of water, the hardness of water, alkalinity of water, buffer intensity, total dissolved oxygen, and inorganic carbon. The indices LSI and LI are discussed. The inhibitor effects of phosphorus and sulfate are discussed. The effect of the presence of metal ions like Zn, Mn, Cu, Cd and Pb on DWDS is discussed. The temperature effect, microbiological induced corrosion and galvanic corrosion processes are briefly presented. The stagnation of water and leaching of Cu, Zn in commonly used DWDS is discussed with the toxicity of iron-based deposits in DWDS. Also, the protection of DWDS from copper pitting is presented. Some of the protection methods to reduced corrosion in steel pipes, Al and Mg-based alloys are also discussed.

Keywords: Corrosion, Drinking Water Distribution Systems, Toxicity Effect.

1. INTRODUCTION

Corrosion is a series of chemical reactions of metals or alloys with their surrounding environment. The reactions of these materials present in the drinking water distribution system such as iron, steel, or their alloys to the different environmental conditions like temperature, wetness, dryness and the characteristics of water pH, salinity, acidic nature, microbiological activity,

---

* **Corresponding author N. Suresh Kumar:** Department of Physics, JNTU College of Engineering, Anantapur-515002, Andhra Pradesh, India; Tel: +91-81211 27157; E-mail:sureshmsc6@gmail.com

N. Suresh Kumar, P. Banerjee, H. Manjunatha and K. Chandra Babu Naidu (Eds.)

presence of heavy metals such as lead, cadmium, arsenic, *etc.*, lead to the corrosion of drinking water systems. Sometimes, these contaminants get dissolved in water leading to health disorders. This results in the change of water quality such as its color, odour, pH, total dissolved solids (TDS), dissolved oxygen (DO), increase in microbiological activities, *etc.* As most diseases are mainly water-borne, it is essential to study the corrosion and corrosion protection in common drinking water systems to provide healthy and pure drinking water to mankind. The corrosion in these pipes can lead to a decrease in the lifetime of supply pipes, a reduction in the supply capacity, and loss of water quality. Corrosion will be present, where two dissimilar metals are contacted under suitable conditions, either directly or indirectly. The corrosion in drinking water distribution systems generally depends on the types of metallic materials used, the quality of water supplied in them, the temperature of the environment, *etc.*

This corrosion will reduce the quality of drinking water to a major extent. The corrosion in DWDS mainly depends on the rate of the flow of water, acidity and alkalinity, the temperature of the water, stagnation of water in the supplied water in the pipes, dissolved O_2, CO_2, presence of rust, sand, sediments, byproducts, salts of chlorides and dissolved sulfates. In addition to this, pitting corrosion also takes place due to the defects in metal pipes. Cast iron and steel pipes occupy a major part in the world's largest supply of drinking water distribution. The main chemical reaction regarding corrosion of cast iron pipes is the conversion of metallic iron (Fe) into ferrous to ferric hydroxides with the reaction of O_2. This results in the release of "red water" finally. The corrosion of iron is quite complex in nature and leads to scale formation, byproduct release, turbidity, and staining. The areas with different concentrations of oxygen and metals can lead to the acceleration of corrosion by biological things [1 - 6]. In this chapter, we would like to present the various factors causing corrosion in DWDS and methods to protect them from corrosion.

2. EFFECT OF PH ON THE CORROSION OF DWDS

The pH of water should be monitored at every stage of drinking water distribution system. If the pH value is small (<3.0), the corrosion becomes high. However, the pH value should be less than 8, as greater pH leads to irritation of the eyes of humans and increases the contamination of chlorine [6]. Increase in pH value leads to weight loss as well as reduced iron concentration levels [7]. The metals like Fe, Pb, Al increase the rate of corrosion when pH of water in DWDS is low [8]. Drinking water parameters such as pH, chloride and sulfate concentrations can provide a measure of corrosivity of the drinking water in the distribution system. Many metals such as lead, iron and aluminium exhibit an increase in corrosion rates in low-pH (less than 5) solutions, when pH is greater than 5, protective layer is formed, which reduces the corrosion due to these metals [9].

3. ALKALINITY AND LANGELIER SATURATION INDEX (LSI)

The mild steel increases the corrosion rates due to chloride and sulphate ions, where the pH value is very low. But in this case, pitting is observed due to the presence of oxygen ions. The normal value of alkalinity of drinking water systems is in the range 100 – 200 mg/L as calcium carbonate [9]. Red water is observed when alkalinity is greater than 60 mg/L as $CaCO_3$ [10]. As the drinking water distribution systems are exposed to environmental effects such as air, pitting is observed under these areas. The presence of oxygen increases the rate of corrosion in this case. But there is no significant effect of calcium or carbonate ions compared to sulphate and chloride salts in water. As the alkalinity is much greater than the sulphate and chloride ions, pitting effect is more within the pH values of 6.5-7 and less once the pH values lie between 8 to 9. As the alkalinity ratio is less than 5, the corrosion rates are increased in drinking water. Low alkalinity and high pH of water leads to the copper pitting corrosion. Larson (1971) stated that the stability of saturation by calcium carbonate is a measure of the quality of water. The Langelier Index can be used to determine this mechanism [10]. We can observe the formation of a protective layer on the pipes, as the water having optimum levels of calcium carbonate indicates a positive Langelier Index. This phenomenon is reversed for a negative Langelier Index, and hence corrion rate is accelerated [11].

4. BUFFER INTENSITY

It is also known as buffer capacity. When the intensity of buffer is high at a constant temperature, the alkalinity decreases the corrosion rate of mild steel in the pH range (6-9) [12]. The loss for cast iron occurred once the buffer intensity reaches a pH 8.4. This is occurred due to the increase in the buffer intensity and further reducing the changes in pH due to corrosion reactions. But at the higher ionic strength and increased conductivity in buffer intensity at different pH levels, the corrosion rates are not decreased [13].

5. TOTAL DISSOLVED INORGANIC CARBON AND ORGANIC CARBON

The presence of inorganic carbon as carbonates, bicarbonates, CO_2, and carbonic acid leads to the formation of carbon films on the surface of the metals. This reduces the corrosion rate. Most of the total alkalinity depends on the total dissolved inorganic carbon in water. The presence of DIC (Dissolved Inorganic Carbonates) decreases the corrosion effect on the copper surface. Also, DIC reduces the iron solubility within the pipe scale. The presence of carbon in the organic species and organic matter will react with piping materials to form a protective layer on the surface of the water pipes. They react with corrosion products leading to the increase in corrosion rates [14]. When total dissolved organic carbon is greater than 3 mg/L, it produces high-risk trihalomethanes (THMs) with free chlorine ions in lead service lines. Natural Organic Matter (NOM) and Aquatic Human Substances (AHS) are removed from the water before supplying in lead service lines [15].

6. DISSOLVED OXYGEN (DO)

DO plays an important role in corrosion. Oxygen is an acceptor of electron to convert Fe^{+2} into Fe^{+3} ion. So, it is obviously clear that DO increases the rate of corrosion in water pipes. Also, the tuberculation which is due to excessive scale or rust build-up in the water pipes, causing the increase of iron with the type of the scale formed in the water pipes [16]. Hence, DO is an effective corrosion agent. Also, in water with DO < 1 mg/L, solutions with phosphates as buffer ions contain a higher corrosion rate compared to water with no phosphates and corrosion rate will be decreased once DO > 1 mg/L. Internal corrosion of water pipes will be controlled by reducing the concentration of DO. Dark brown or light blue colored corrosion products are observed with an uncontrolled concentration of DO on the surface of pipes [17]. The change in the DO is highly relative to the release of Fe in water pipe systems [18].

7. HARDNESS OF WATER

The hardness is defined as the concentration of dissolved calcium and magnesium salts present in the water (soft-hard 60 mg/L- highly hard180 mg/L). As the concentration of these salts is high, water is hard which reduces the corrosion rate by forming thin protective layers on the surface of water pipes depending on various factors like pH, alkalinity, turbidity, and other salts. Herein, the soft water with fewer concentrations of calcium and magnesium salts increases the corrosion rate. Soft water supports uniform corrosion while hard water supports corrosion by hard deposits [19].

8. LARSON INDEX (LI)

The ratio of chloride and sulfate to bicarbonate expressed in the Larson Index:

$$\text{Larson Index} = \frac{2[SO_4^{-2}] + [Cl^-]}{[HCO_3^-]} \qquad (1)$$

If LI is high, the water is more corrosive in nature. The presence of sulfate and chlorine leads to an increase the iron concentration. Higher chloride to sulfate ratio increases the corrosion. But in contrary sulfate prevents the dissolution of iron oxides leading to decrease in iron concentrations reported in literature [20]. The sulfate and chloride in drinking water distribution systems can cause the formation of a protective scale on steel surfaces reducing the corrosion effect by absorbing iron in the inner walls of PVC-U water pipes.

9. PHOSPHATE AND SILICATE-BASED INHIBITOR EFFECTS

Phosphate or silicate-based inhibitors are added to drinking water from the early 20th century. Since phosphate can reduce the red water problem and corrosion of cast iron pipes, phosphate inhibitors are generally used to reduce the precipitation of calcite growth in water pipes. The common types of phosphate inhibitors used are hexametaphosphate (example $Na_{22}P_{20}O_{61}$), orthophosphate, zinc metaphosphate, zinc orthophosphate, and bimetallic phosphate (sodium-zinc- or potassium-zinc phosphate) *etc*. Phosphate inhibitors completely do not reduce iron but lower the rate of corrosion. These phosphate inhibitors convert lead, copper, manganese and further reduce the quality of water. The addition of phosphate inhibitors reduces the iron corrosion by 5 times in galvanized pipes and corrosion by Cu by 2 times forming the Cu_2O layer [21].

Sodium silicate is primarily used for the control of corrosion in different types of metallic drinking water distribution pipes. Silicates reduce the corrosion rate by forming a protective film on the surface of metallic pipes. The sodium silicate is used for the pH > 6.0 water in the DWDS. When water contains more chloride, more silicate is to be used for disinfecting water. The negative charge on silica moves to anode, and most of the negative charge is neutralized. Also, the silicate films formed due to the addition of silicate are electrically insulator. It is beneficial for several applications. But the amount of effective reduction of corrosion rate is to be identified for efficient water distribution. The formation of films on Zn-based pipes is given by Lehrman and Shuldener after adding sodium silicate to hot water continuously for a year. Therefore, the film is unevenly distributed by different compositions in brass, SiO_2 plus copper oxide film and in galvanized pipes while the evenly distributed ZnO film is observed. The precipitates of Fe_2O_3 and organic matter are commonly observed [22, 23].

10. LEAD AND COPPER RULE (LCR)

The presence of lead causes a delay in the physical and mental development of newly born children. The copper shows an impact on other gastrointestinal problems including damage to liver and kidneys. According to the lead and copper rule (LCR, EPA-1991), water distribution systems are required to have corrosion control measures. Also, the optimal levels of lead and copper in the drinking water systems given as 0.015 mg/L for lead and 1.3 mg/L for copper can reduce the risk of lead and copper in drinking water systems.

11. PRESENCE OF CADMIUM AND ZINC

The cadmium in drinking water is due to natural erosion of the cadmium as well as cement pipelining of drinking water systems. The presence of cadmium in the water causes diarrhea, nausea, muscle cramps, liver damage, renal problems *etc* [24]. But as the age of pipes is increasing, the risk of lead and cadmium is decreases. Zinc in drinking water is also formed due to natural erosion. It is found that the galvanized pipes coated with zinc contain traces of major impurities of lead, copper, chromium, barium, aluminium and other corrosive impurities. Even though zinc is essential for humans for growth and development [25], and the higher concentrations of zinc lead to nausea, vomiting and abdominal cramps [26]. The contamination of water with zinc adds metallic taste to water [27]. The limit for zinc is 5 mg/L by USEPA.

12. TEMPERATURE EFFECT

The effect of temperature on corrosion of pipes is more at high temperatures than at low temperatures. Due to the variation of seasonal temperatures, the water pipes buried will experience different temperature during the water supply. Fiksdal studied that the decrease in weight loss observed for iron samples at 13 and 20°C. Volk *et al.* [9] found lower iron concentrations as well as low corrosion rates. During the cold winter, there is a possibility of red water. The biological activities DO, viscosity, iron conversion into iron hydroxide depends on the temperature of the pipes. The microbiological activity is increased due to an increase in the temperatures. At higher temperatures, the oxidation of iron is increased due to an increase in biological activity. At high temperatures, the availability of oxygen is less, and hence, the corrosion is limited. The Kuch reaction converts ferric iron scale and acts as the electron acceptor. So, ferrous metal is formed due to the oxidation of iron (Kuch, 1988). If the diffusion of iron is limited, the increase in the temperature of pipes decreases the viscosity of water. At this moment, the corrosion is increased as DO, and acceptors of electrons are increased. The rate of oxidation of Fe^{+2} is increased at a constant pH for each 15°C (Millero, 1989; Stumm, Morgan, 1996). This leads to the formation of different compounds at different temperatures. Then, the corrosion depends on the different scale formed in the pipes.

13. MICROBIOLOGICAL ACTIVITIES

Disinfection of water before the water transportation in the drinking water distribution system is mandatory, and it is governed by various laws. Scientists studied that the microorganisms grow 10^4-10^6 cells per liter, including various bacteria, live amoebae, fungi as well as viruses. The disinfection process is almost completed with the chlorination process before distribution. This significantly reduces the microorganisms. At the point of use, water in DWDS is to be at most free from infectious bacteria and other viruses. Increased turbidity and large particle counts induce the microbial growth in water [28]. Hence, at the point of use, more microorganisms are observed than the water leaving from the treatment plant [29].

The corrosion due to microorganisms is known as Microorganisms Induced Corrosion (MIC). The microorganisms grow at the stagnated water in the pipes and where the pipes are deposited. Reducing the pH of water, the with acidic micro-organisms corrosion processes will be increased. The hydrogen sulfide created due to the low pH increases the corrosion of metal pipes [30]. The nitrate-reducing bacteria is found to be increasing the corrosion rates. The microorganisms causing corrosion is different at a different location around the world. The use of chlorine for disinfecting microorganisms is a clear indication that the MIC is also important in DWDS. Stainless steel and cast-iron pipes can also observe the MIC failure. The copper coating and other anti-bacterial plating methods may be used to reduce MIC in DWDS. But complete elimination of MIC is almost not possible.

14. PROTECTION OF DWDS

14.1. Effect of Chloramination

The pipe wall biofilm is the majority part of the biomass in the water distribution systems. The removal of bacteria grown on this film is very difficult when compared to the suspended bacterial growth in water. The chlorination cannot completely reduce the growth of the biofilm in plastic water pipe systems [31]. The disinfection methods used for the case of cast iron pipes which are a majority water supply system in the world also causing the corrosion process. This includes the formation of iron oxide leading to red water problem at the point of use. The use of chloramines instead of chlorine significantly reduces the bacteria causing the growth of bio-corrosion [32]. The dissolved organic carbon, chlorine and sulphate ions play a vital role in the growth of corrosion biofilm and corrosion bacteria. All these factors are to be considered in the chloramination process. The disinfecting processes will alter the kinetics of corrosion process as well as water quality. The disinfecting process increases the corrosion rates with ClO compared to NaClO in case of cast-iron pipes. At the last stages of the disinfection process, the bacteria can also become a factor for promoting corrosion process. The iron-reducing bacteria and high pH values produce alpha-Fe (OOH) and Fe_2O_3 with the use of NaClO at the second and third stages of disinfecting water. The use of ClO reduces the corrosion rates at the last stage of disinfecting water [33].

14.2. Toxicity of Iron-based Deposits

The cast-iron pipes will have deposits of Fe_3O_4 and Fe_2O_3 during the corrosion process. The Fe_3O_4 and Fe_2O_3 are less toxic compared to metal oxides such as ZnO, and CuO *etc*., when they are present in the water sources in the drinking water systems. However, as the particle size is reduced, their toxicity becomes more. The FeOOH particles grown on different materials like inorganic and organic have different toxicity nature. The particles consisting of sharp edges will be more reactive for corrosion since they are biologically and chemically active [34]. The FeOOH nanoparticles with urchin-like nanostructures are potentially toxic. These FeOOH nanoparticles generally accumulate at the liver and spleen causing the damage of cells. The interaction between the FeOOH and perfluorooctanoic acid (PFOA) contains the highest toxicity effect even it does not have sharp surfaces compared to FeOOH nanostructures grown on Al, Ca, Mn, Si, bisphenol A (BPA) [35].

14.3. Galvanic Corrosion Between Stainless-Steel and Lead

The permitted level of lead in drinking water distribution varies from $10 - 15$ µg/l by various environmental protection agencies like EU, WHO and USEPA. When lead pipes from aged water distribution connected to stainless steel pipes, the lead particulates are formed after twelve weeks. The formation of lead is more, as connections are made without galvanic connection. This is due to Pb (II) ions released from the lead carbonates and lead sulfates. The galvanic corrosion increases the release of Pb, which accelerates the deposition corrosion. Higher CSMR and pH increase the Pb release in tap waters tested [36, 37].

14.4. Protection from Galvanic Corrosion Between Steel and Lead

This can be reduced by replacing lead (Pb) connections with steel with other less corrosive metals. Also, the addition of inhibitors like orthophosphate decreases the release of Pb from galvanic corrosion.

14.5. The Effect of Stagnation Time and Temperature of Water in Leaching of Copper and Zinc in DWDS

The stagnation time and temperature of the water play an important role in corrosion processes and microbiocidal activity. The effect of temperature on iron corrosion is often overlooked. Many parameters such as the dissolved oxygen (DO), turbidity, viscosity, oxidation rate of iron, the biological activity, the scale formation, and many other temperature dependent thermodynamic properties of water in DWDS. Due to the stagnation of water in DWDS, the copper concentration is increased. The maximum Cu leaching is found to be more in winter and more in the summer season. The formation of free copper causes a reaction with the organic matter leading to the formation of water-soluble products and precipitates. This leads to an increase in the corrosion rates by copper leaching. But after reaching the peak values of the release of copper, the Cu is converted into CuO with reacting with the dissolved oxygen present in the drinking water distribution systems. There are more studies indicating the leaching of Zn from brass pipes in drinking water distribution systems. The leaching of Zn is also decreased like Cu, due to a decrease in the dissolved oxygen levels in stagnant water in drinking water distribution systems. Later the corrosive products were observed on the surface of water pipes which accelerates the corrosion process. The total organic carbon is also found to be decreased in Cu based DWDS. This is large in the winter season than in summer. The reason is bio-scales were developed on the copper surfaces where the bacteria can use some part of the inorganic carbon during the winter season. So, the decrease in the temperature of the water during the winter searching increases the bacterial activity in DWDS. The HPC bacteria are increased in the winter season. But the other measurements for ATP and flow cytometry indicated that there is an influence of stagnant water both in winter and summer seasons. The formation of Cu during the stagnation water shows a different impact on the toxicity and growth of some bacteria because of resistance developed by the presence of excess copper in DWDS. This is quite complex in nature [38].

14.6. The Corrosive Nature of Mn in DWDs and Protection of Pipes from Mn

The presence of manganese in drinking water causes the problem of black water, stain and dirt to water supplied like the problem of red water created by the presence of iron. These two elements cause the color to the water. In humans, it can cause a problem called magnesium, like Parkinson's disease. Manganese is present in the form of soluble manganese in the water. The Hydraulic systems which use PVC and iron pipes have deposits of Mn [39]. The allowed levels of manganese are 0.05 to 0.1 µg/l by USEPA and WHO, respectively. The PVC and PACK pipes are having significant amounts of Mn in them. The Langelier, Riznar, Mojmir indices confirmed the corrosive nature of Mn and other solids present in the water reported by Alavrez-Bastida in the pipes of Latin America [40, 41]. Generally, during the corrosion process, the Mn is liberated from the source since it is not found in the water collected from the source. But traces are found in tap water. The Mn generally sticks on the PVC and iron pipes. They can be removed with the pressure of water in drinking water distribution systems. The Mn can be removed using modified zeolitic stuff. The removal efficiency of Mn is higher in aqueous solutions compared to drinking water.

14.7. Corrosion Protection of Steel

The protective nature of N,N'dimethylaminoethanol (DMEA) due to chloride ions was reported in literature. The N,N'dimethylaminoethanol (DMEA) is an organic inhibitor, and it can be used for the protection of rebars. The electrochemical reaction contaminated with the chloride ions can cause corrosion of steel bars. The corrosion rate of steel is increased with the increase of chlorine ions. Chloride ions are responsible for the corrosion of the bars by destroying the passive films formed on the rebar surfaces constructed. According to their study, the addition of 2 wt % DMEA reduces the corrosion rate by 43% of the steel bars. Also, the addition of DMEA inhibitor reduces the pitting of the area of the corrosion of steel bars. The $Ca(OH)_2$ solution with different amounts of chloride ions indicates the increase in the corrosion process in the absence of the addition of the inhibitor DMEA as well lowering of the pitting potentials [42].

14.8. Copper Pitting in DWDs and Protection

Copper corrosion results from the loss of copper metal to the solution. There are several reasons for pitting of copper corrosion in water having the pH, dissolved oxygen, alkalinity, presence of chloride ions, sulfate, phosphate, organic matter such as inorganic carbon (IOC), stagnation of water and temperature of water in

DWDS. The pinhole leaks (which can be observed on all portions of pipes is associated with copper) are more for the case of cold water compared to the hot water. In copper pitting high pH value, low alkalinity and dominated levels of chloride and sulfate ions were identified. The pit structure contains pit cap, pit perforated membrane and pit. The pit caps are associated with sulfates. The membrane lies between pit cap and pit is brittle having a thickness of the order of micrometers, porous, and non-continuous in the region which covers the pit, and it contains cuprite mineral. Hot water is associated with corrosion by forming the copper by-products. The CuO is generally observed in hot water pipes as a by-product [43].

14.9. Protection of Al Based Pipes from Corrosion

Aluminium alloys are widely used in many industries, including water pipes systems because of their low cost, light weight, and strength properties. But the surface of these alloys easily corroded due to chemical reactions. The polymer coatings on these pipes are the simple methods to protect the metal from corrosion. Copolymerization and electro polymerization are some of the methods used to reduce the corrosion of aluminium based alloys due to their simple and low-cost approaches. Chitosan (CS), O-Phenylenediamine(o-PDA) monomers are used in copolymerization process. The performance of these monomers can be increased by using high valence lanthanum (La) and zirconium (Zr) by increasing the coordination with polymer [44]. The use of Zr is increasing attention because of its low toxicity, non-reactive nature, high thermal stability, less reactive to biological activity. Also, the high positive charge on Zr attracts the negative ions, as well as Zr. This can be used to remove the pollutants, and it has less leaching of water. Optimized polymers of oPDA and CS are coated by electrosynthesis method on Al metal. The Zr is coated on the oPDA and CS coatings by applying dc-voltage. These coatings are studied for fluorine removal and testing of corrosion rates. Still, 35% of the population of the world depends on groundwater, and the peculiar problem with its consumption is fluoride. The Zr coatings on polymerized oPDA and CS can remove the fluorine by complexation and absorption process. This is an endothermic and spontaneous process. The Zr based polymer coating also acts as a sorbent for other ions like sulphate, chlorine, nitrate, and bicarbonate ions. Among these ions, sulphate ions are strongly attracted by the Zr coating rather than the other ions. Also, the coatings of Zr loaded copolymerization have an excellent result on the reduction of fluorine in water and decreasing the corrosion of Al based alloys. As this method is very simple, easy, and biocompatible, it can be used for large scale for the protection of DWDS.

14.10. Protection of Magnesium Based Alloys from Corrosion

Magnesium alloys are commonly used, when there is a need for light weight, and strength. Among the Mg alloys, the most used alloy is AZ91. But the common problem regarding magnesium alloys is the insufficient corrosion resistance and highly reactive nature to the environment. The corrosion of the Mg alloys is commonly reduced significantly by using sol-gel coatings. These sol-gel coatings are prepared based on silanes-amino acids along with the addition of some nanoparticles like TiO_2, *etc* [45, 46]. The organic compounds with special properties can be used as inhibitors to protect from the corrosion. The commonly used organic inhibitors have aromatic rings in their structure. They are generally the compounds of nitrogen, oxygen, sulphur, *etc*. Amino acids are environmentally friendly, non-toxic, and cheap. They can be prepared with high purity. L-methionine, cysteine (CYS), serine (SER), aminobutyric (ABU) acid, threonine (THR), alanine (ALA), valine (VAL), s- phenylalanine (PHE), tryptophan (TRP) and tyrosine (TYR) are some of them. Silica films based on TEOS and MTES are developed by Sorkhabi *et al.* [45] to protect magnesium from corrosion without any cracks in them. These coatings are added with amino acids. The presence of amino acids in the coatings reduces the corrosion consistently. The coatings coated with amino acids are crack free. The alloys coated with 1 wt % of L-Aspartic acid show maximum corrosion resistance. L-Methionine and L-Glutamine exhibit week corrosion protection compared to L-Aspartic acid.

Ashraf *et al.* [46] showed that the hybrid sol-gel coatings with different amino acids as an inhibitor and TiO_2 nanoparticles. The corrosion protection of the coatings is in the sequence of the amino acids are in the order of cysteine > serine > alanine > arginine. When TiO_2 nanoparticles are added with amino acids, it can significantly improve the corrosion protective nature. A sol-gel coating is prepared from mixing tetraethylorthosilicate (TEOS) as inorganic part. The triethoxyvinylsilane (TEVS) coated with 0.5 wt % cysteine and 1.0 wt % TiO_2 nanoparticles can protect the AZ91 magnesium alloy in NaCl solutions from corrosion for a longer duration of exposure.

CONCLUSION

The corrosion in drinking water distribution system is mainly due to various factors like pH of water, the hardness of water, alkalinity, presence of metal ions and toxic elements, leaching of metals, galvanic corrosion *etc*. The protection of DWDS from the various corrosion methods can be done by appropriate methods depending on the type of corrosion. The effect of MIC and its protection is discussed. The protection of DWDS by sol-gel coatings with amino acids, chloramination, co-polymerization using monomers is also discussed.

CONSENT FOR PUBLICATION

Not applicable.

CONFLICT OF INTEREST

The author declares no conflict of interest, financial or otherwise.

ACKNOWLEDGEMENTS

Declared none.

REFERENCES

[1] Gedge, G. Corrosion of Cast Iron in Potable Water Service. *Proceedings of the Institute of Materials Conference,* London, UK**1992**.

[2] AWWA. AWWA Report: $325 billion for Pipes *AWWA Mainstream,* **1999**.

[3] Davies, C.; Fraser, D.L.; Hertzler, P.C.; Jones, R.T. *USEPA's Infrastructure Needs Survey Journal AWWA,* **1997**.

[4] American Water Works Association. American Water Works Association. *WATER:\STATS 1996 Survey,* **1996**.

[5] Bray, A.V. *Personal communication.,* **1997**.

[6] WHO Working Group. Health impact of acidic deposition. *Sci. Total Environ.,* **1986**, *52*(3), 157-187.
[http://dx.doi.org/10.1016/0048-9697(86)90118-X] [PMID: 3738500]

[7] Kashinkunti, R.D.; Metz, D.H.; Hartman, D.J.; DeMarco, J. How to reduce lead corrosion without increasing iron release in the distribution system. *Proc. AWWA Water Quality Technology Conference,* Tampa, FL**1999**.

[8] Ding-Quan, N.G.; Yi-Pin, L.; Hartman, D.J.; DeMarco, J. Effects of pH value, chloride and sulfate concentrationson galvanic corrosion between lead and copper in drinking water. *Environ. Chem.,* **2016**, *13*, 602-610.
[http://dx.doi.org/10.1071/EN15156]

[9] Christian, V.; Esther, D.; John, S.; Mark, L. Practical evaluation of iron corrosion control in a drinking water distributionsystem wat. *res,* **2000**, *34*(6), 1967-1974.

[10] Horsley, M.B.; Northrup, B.W.; O'Brien, W.J.; Harms, L.L. Minimizing iron corrosion in lime softened water. *Proc. AWWA Water Quality Technology Conference,* , pp. 5C-3.**1998**

[11] Larson, T.E. *Corrosion Phenomena-Causes and cures,* 3rd ed; McGraw-Hill Book Co.: New York, **1971**, pp. 295-312.

[12] Langelier, W.F. The analytical control of anti-corrosion water treatment. *J. Am. Water Works Assoc.,* **1936**, *28*, 1500-1521.
[http://dx.doi.org/10.1002/j.1551-8833.1936.tb13785.x]

[13] Rodolfo, A.; Jr, Pisigan Edward singleyinfluence of buffer capacity, chlorine residual, and flow rate on corrosion of mild steel and copper. *J American Water Works Association,* **1987**, *79*(2), 62-70.

[14] Brian, R.L.; Kathie, L.L.; Zorabel, M. Garno Impact of pH, Dissolved Inorganic Carbon, and

Polyphosphates for the Initial Stages of Water Corrosion of Copper Surfaces Investigated by AFM and NEXAFS
[http://dx.doi.org/10.5618/chem.2011.v1.n1.3]

[15]   Lisa, D.W.; Beata, G.; Kenneth, B. Effect of total organic carbon and aquatic humic substances on the occurrence of lead at the tapWater. *Qual. Res. J.,* **2017**, *52*(1), 2-10.

[16]   Baylis, J.R. *Cast-Iron Pipe Coatings and Corrosion.,* **1953**.
[http://dx.doi.org/10.1002/j.1551-8833.1953.tb16493.x]

[17]   Jung, Haeryong; Kim, Unji; Lee, Hyundong Effect of Dissolved Oxygen (DO) on Internal Corrosion of Water Pipes" Environmental Engineering Research **2009**, *14*(3), 195-199.

[18]   Hu, J.; Dong, H.; Xu, Q.; Ling, W.; Qu, J.; Qiang, Z. Impacts of water quality on the corrosion of cast iron pipes for water distribution and proposed source water switch strategy. *Water Res.,* **2018**, *129*, 428-435.
[http://dx.doi.org/10.1016/j.watres.2017.10.065] [PMID: 29179122]

[19]   VigneshNallasivam, Elayaperumal "Comparison of corrosion characteristics of carbon steel in soft and hard waters under different parameters" Conference Paper No. 17030 The 17th Asian Pacific Corrosion Control Conference, January 27-30, 2016, IIT Bombay, Mumbai, India.

[20]   Wang, Jiaying; Tao, Tao *Public Health,* **2017**, *14*(6), 660.
[PMID: 29072854]

[21]   Ascott, M.J.; Gooddy, D.C.; Lapworth, D.J.; Stuart, M.E. Estimating the leakage contribution of phosphate dosed drinking water to environmental phosphorus pollution at the national-scale. *Sci. Total Environ.,* **2016**, *572*, 1534-1542.
[http://dx.doi.org/10.1016/j.scitotenv.2015.12.121] [PMID: 26774133]

[22]   Lehrman, L.; Shuldener, H.L. The role of sodium silicate in inhibiting corrosion by film formation on water piping. *J. Am. Water Works Assoc.,* **1951**, *43*, 175.
[http://dx.doi.org/10.1002/j.1551-8833.1951.tb15225.x]

[23]   Lehrman, L.; Shuldener, H.L. Action of sodium silicate as a corrosion inhibitor in water piping. *Ind. Eng. Chem.,* **1952**, *44*, 1765.

[24]   Lauwerys, R.R. Health effects of Cadmium. In: *Trace metal: Exposure and Health Effects*; Di Ferrante, E., Ed.; Pergamon Press: Oxford, England, **1979**; pp. 43-64.

[25]   Laura, M. Plum, Lothar Rink, and HajoHaaseThe Essential Toxin: Impact of Zinc on Human Health Int J Environ Res. *Public Health,* **2010**, *7*(4), 1342-1365.

[26]   Zinc in Drinking-water WHO/SDE/WSH/03.04/17.

[27]   Potgieter, S.; Pinto, A.; Sigudu, M.; du Preez, H.; Ncube, E.; Venter, S. Long-term spatial and temporal microbial community dynamics in a large-scale drinking water distribution system with multiple disinfectant regimes. *Water Res.,* **2018**, *139*, 406-419.
[http://dx.doi.org/10.1016/j.watres.2018.03.077] [PMID: 29673939]

[28]   Hammes, F.; Berney, M.; Wang, Y.; Vital, M.; Köster, O.; Egli, T. Flow-cytometric total bacterial cell counts as a descriptive microbiological parameter for drinking water treatment processes. *Water Res.,* **2008**, *42*(1-2), 269-277.
[http://dx.doi.org/10.1016/j.watres.2007.07.009] [PMID: 17659762]

[29]   van der Wielen, J.; Bakker, G.; Atsma, A.; Lut, M.; Roeselers, G.; Graaf, B.A. Survey of indicator parameters to monitor regrowth in unchlorinated drinking water. *Environ. Sci. Water Res. Technol.,* **2016**, *2*(4), 683-692.

[30]   Mand, J.; Park, H.S.; Jack, T.R.; Voordouw, G. The role of acetogens in microbially influenced corrosion of steel. *Front. Microbiol.,* **2014**, *5*, 268.
[http://dx.doi.org/10.3389/fmicb.2014.00268] [PMID: 24917861]

[31]   Gomes, I.B.; Simões, M.; Simões, L.C. The effects of sodium hypochlorite against selected drinking

water-isolated bacteria in planktonic and sessile states. *Sci. Total Environ.,* **2016**, *565*, 40-48.
[http://dx.doi.org/10.1016/j.scitotenv.2016.04.136] [PMID: 27156214]

[32]   Wang, H.; Hu, C.; Hu, X.; Yang, M.; Qu, J. Effects of disinfectant and biofilm on the corrosion of cast iron pipes in a reclaimed water distribution system. *Water Res.,* **2012**, *46*(4), 1070-1078.
[http://dx.doi.org/10.1016/j.watres.2011.12.001] [PMID: 22209261]

[33]   Zhang, H.; Tian, Y.; Kang, M.; Chen, C.; Song, Y.; Li, H. Effects of chlorination/chlorine dioxide disinfection on biofilm bacterial community and corrosion process in a reclaimed water distribution system. *Chemosphere,* **2019**, *215*, 62-73.
[http://dx.doi.org/10.1016/j.chemosphere.2018.09.181] [PMID: 30312918]

[34]   Zhuang, Y.; Han, B.; Chen, R.; Shi, B. Structural transformation and potential toxicity of iron-based deposits in drinking water distribution systems. *Water Res.,* **2019**, *165*, 114999.
[http://dx.doi.org/10.1016/j.watres.2019.114999] [PMID: 31465995]

[35]   Zhuang, Y.; Kong, Y.; Liu, Q.Z.; Shi, B.Y. Alcohol-assisted self-assembled 3D hierarchical iron hydroxide nanostructures for water treatment. *CrystEngComm,* **2017**, *19*(39), 5926-5933.
[http://dx.doi.org/10.1039/C7CE01320E]

[36]   Ng, D.Q.; Chen, C-Y.; Lin, Y.P. A new scenario of lead contamination in potable water distribution systems: Galvanic corrosion between lead and stainless steel. *Sci. Total Environ.,* **2018**, *637-638*, 1423-1431.
[http://dx.doi.org/10.1016/j.scitotenv.2018.05.114] [PMID: 29801235]

[37]   ZhangY.Dezincification and brass lead leaching in premise plumbing systems: effect of alloy, physical conditions and water chemistry, master thesis. Virginia, USA, **2009**. vtechworks.lib.vt.edu

[38]   Lj, System; J.P., Zlatanovi; der Hoek, Van An experimental study on the influence of water stagnation and temperature change on water quality in a full-scale domestic drinking water Research **2017**, *123*, 761-772.

[39]   Cerrato, J.M.; Reyes, L.P.; Alvarado, C.N.; Dietrich, A.M. Effect of PVC and iron materials on Mn(II) deposition in drinking water distribution systems. *Water Res.,* **2006**, *40*(14), 2720-2726.
[http://dx.doi.org/10.1016/j.watres.2006.04.035] [PMID: 16765409]

[40]   Alvarez-Bastida, C.; Martínez-Miranda, V.; Vázquez-Mejía, G.; Solache-Ríos, M.; Fonseca-Montes de Oca, G.; Trujillo-Flores, E. The corrosive nature of manganese in drinking water. *Sci. Total Environ.,* **2013**, *447*, 10-16.
[http://dx.doi.org/10.1016/j.scitotenv.2013.01.005] [PMID: 23376288]

[41]   Alvarez-Bastida, C. V.MartinezMiranda, M.SolacheRiosI.Linares Hernandez, A.Teutli-Sequeira, G.Vazquez Mejia, Drinking water characterization and removal of manganese. Removal of manganese from water. *J. Environ. Chem. Eng.,* **2018**, *6*(2), 2119-2125.
[http://dx.doi.org/10.1016/j.jece.2018.03.019]

[42]   Rakanta, E.; Zafeiropoulou, Th.; Batis, G. Corrosion protection of steel with DMEA-based organic inhibitor. *Constr. Build. Mater.,* **2013**, *44*, 507-513.
[http://dx.doi.org/10.1016/j.conbuildmat.2013.03.030]

[43]   Darren, A. Lytle, Mallikarjuna N. Nadagouda A comprehensive investigation of copper pitting corrosion in a drinking water distribution system. *Corros. Sci.,* **2010**, *52*, 1927-1938.
[http://dx.doi.org/10.1016/j.corsci.2010.02.013]

[44]   Mohan Raj, R. V. Raj Electrosynthesis of Zr-loaded copolymer coatings on Al for defluorination of water and its corrosion protection ability. *Prog. Org. Coat.,* **2019**, *137*, 105065.
[http://dx.doi.org/10.1016/j.porgcoat.2019.04.039]

[45]   Ashassi-Sorkhabia, Habib; Moradi-Alavian, Saleh; D., Mehdi Hybrid sol-gel coatings based on silanes-amino acids for corrosion protection of AZ91 magnesium alloy: Electrochemical and DFT insights Progresses in Organic Coatings **2009**, *131*, 191-202.

[46] Ashraf, M.A.; Liu, Z.; Peng, W-X. Nasser Yoysefi Amino acid and TiO$_2$ nanoparticles mixture inserted into sol-gel coatings: An efficient corrosion protection system for AZ91 magnesium alloy. *Prog. Org. Coat.,* **2019**, *136*, 105296.
[http://dx.doi.org/10.1016/j.porgcoat.2019.105296]

CHAPTER 6

Corrosion in Reinforcement Cement Concrete

B. Venkata Shiva Reddy[1, 2], N. Suresh Kumar[3], K. Chandra Babu Naidu[1,*], D. Baba Basha[4], M. Balaraju[5] and T. Anil Babu[1]

[1] *Department of Physics, GITAM Deemed to be University, Bangalore - 562163, Karnataka, India*

[2] *Department of Physics, The National College, Bagepalli-561207, Karnataka, India*

[3] *Department of Physics, JNTU College of Engineering Anantapur, Anantapuramu-515002, A.P., India*

[4] *Department of Physics, College of Computer and Information Sciences, Majmaah University, Al'Majmaah, Saudi Arabia*

[5] *Department of Chemistry, PSC & KVSC Govt. Degree College, Nandyal, Kurnool-518502, A.P., India*

Abstract: Corrosion is an electrochemical reaction initiated by many factors like the ingress of chloride particles and carbon particles. The electrons can move in the steel rebar and the ions can move in the concrete, which acts as an electrode leading to corrosion. The rate of corrosion can be mitigated by the addition of corrosion inhibitors into the concrete. The corrosion increases the volume of rebar and hence, cracking of concrete takes place. The cracking of concrete can be mitigated by the addition of fibers and bacteria into the concrete. The bacteria can produce calcium carbonate, which helps in the self-healing of cracks in concrete. To examine the damage of the reinforcement in the concrete, X-ray microcomputed tomography is employed without wasting of testing sample.

Keywords: Bacteria, Chloride and Carbon Particles, Corrosion, Electrochemical Reaction.

1. INTRODUCTION

Cement is the general name for powdered materials that initially have plastic flow when mixed with water. But they form a solid structure after several hours with varying degrees of strength and bonding properties. The compositions of cement

---

* **Corresponding author K. Chandra Babu Naidu:** Department of Physics, GITAM Deemed to be University, Bangalore-562163, Karnataka, India; Tel: +91-9398426009; E-mail: chandrababu954@gmail.com

N. Suresh Kumar, P. Banerjee, H. Manjunatha and K. Chandra Babu Naidu (Eds.)

are Al_2O_3, SiO_2, SO_3, K_2O, CaO and Fe_2O_3. The corrosion in reinforcements of cement is a common problem, but a slow process. Corrosion is a process of electrochemical reaction that takes place due to the flow of charges like ions or electrons. The industrial material like steel and iron get corrosion as they are not naturally available materials but occurs in the form of ore. The steel, like many materials except platinum and gold, is unstable under atmospheric conditions and revert into its natural state with the release of energy-iron oxide or rusting. This process is called corrosion. Chloride induction into steel is a common cause for corrosion in reinforcement structure and dissolution of iron. This leads to the formation of iron oxides. The corrosion takes place as the chloride value is greater than the threshold chloride value. However, corrosion initiation depends on many factors like quality of steel or concrete interface, the chemistry of the pore solution, potential of the steel and the orientation of the reinforcement to the concrete casting direction, also including macro factors such as macro pores, air voids and cracks [1]. The corrosion profile reinforcement can be characterized with impressed current densities by X-ray micro-computed tomography (XCT). This is a non-distractive method than any other methods. It is impossible to nondestructively monitor with the process of the gravimetric method and 3D scanning. Even though electrochemical methods (Linear polarization resistance and electrochemical impedance spectroscope) are nondestructively tested, they cannot be used to visualize the morphology of reinforcement. The impressed current density may be varied depending on the morphology of the reinforcement and concrete cracking.

The XCT test is conducted to trace accelerated corrosion with the impressed current. The chemical compositions of reinforcement are C, Si, Mn, P, S, Cr and Fe [2]. In railway engineering, shotcrete (spay concrete) is used in railway tunneling, in which the steel rods are used as a reinforcement. In the electrified railway system, the train gets electric current by overhead connection to the electric line. Thus, in an electric train, the stray current is developed, which disperses from the current return path, such as running track, surrounding buildings and infrastructure. The reinforcement steel in the shotcrete can pick up the stray current, which leads to corrosion [3]. The reinforcement concretes are built-in railway/highway bridges, high rise buildings and power plants with a target life of more than one hundred years. To get such a long target life, the steel is coated with organic inhibitors and coating to improve the corrosive resistance power. The organic coating provides a shield or physical barrier between the under-laying steel and deleterious elements like chloride elements, moisture and oxygen and restricts the ionic and cathodic areas. The cement polymer composite and the acrylic base is a type of organic coating which is widely used in the construction industry [4]. Incorporation of inhibitors (Fly ash or silica fume) into the concrete also improves the performance of concrete. Silica fumes are more costly than fly ash. Silica fume is more efficient pozzolanic material than fly ash. Fly ash is the by-product of industrial waste produced by the burning of coal combustion in electric power generation [5].

The corrosion is of two types, one is general (uniform) and pitting (localized). The most direct cause for corrosion is a reduction in the reinforcement of diameter of material and cross-sectional area. This reduction is responsible for the structural safety and integrity of the reinforced concrete. If the cross-section depreciation is high, the working stress is also maximum in the reinforcement. The corrosion produces insoluble bi-products commonly known as rusting, which increases more volume 3 to 8-fold than that of the original volume of the material used. This expansion leads to cracking, spalling and delamination of the concrete cover and bond loss between steel and concrete. This further accelerates corrosion rate and hence reduces the serviceability of concrete [6]. The corrosion rate is directly proportional to the electrical resistivity and inversely proportional to the concrete resistivity. The presence of micro silica and nano silica additives reduces the corrosion rate in concrete [7]. A new type of cement-based composites coral aggregate composites is being used for specific purposes in ocean-going Iceland reefs projects. The coral aggregates are economical and high efficiency. The coral aggregate replaces the natural sandstone in the preparation of coral aggregate concrete. Coral aggregates are composed of aragonite, dolomite, and calcite. The main ingredient is calcium carbonate and a little bit of chlorinated salt [8]. In sea or ocean, the Cl⁻ decides the durability and strength of the concrete structure of reef. However, the porous structure of coral and Cl⁻ in seawater leads to the corrosion of reinforcement steel. The apparent chloride diffusion coefficient and surface free chloride concentration are important parameters to assess the diffusion of Cl⁻ ions and service life of concrete structure [9]. The coral aggregate concrete is made by locally abundant coral reef debris. This is secreted by algae and coral polyps [10].

The magnesium oxychloride cement concrete [MOCC] is an improved form of cement that is ecofriendly and reduces waste resources. This concrete can exhibit the improved energy conservation and reduction of emission. The MOCC is a good halogen resistor without modification and the chemical composition is $MgO\text{-}MgCl_2\text{-}H_2O$. When steel starts corroding, the dimensional changes can take place in the steel *i.e.*, the cross-sectional area is decreased, the bond between concrete and steel reduces the formation of cracks. Hence, it leads the structural damage. The following methods can be employed for the investigation of steel corrosion:

- Dry and wet cycle test,
- Chloride ion penetration test,

• Addition of chloride salt to concrete.

However, the above tests are time-consuming which can take years to obtain the results. Hence, to get quick results, we can follow the galvanometric accelerated test techniques as follows:

• Scanning electron microscope,
• X-ray diffraction technique,
• Energy dispersive X-ray spectroscope.

However, these tests are also destructive because we must grind and slice the samples. In addition, during grinding process, the microstructure will be destroyed. Hence, the latest method is X-ray computed tomography that is being employed in corrosion tests [11]. The schematic representation of X-ray computed tomography is shown in Fig. (**1**).

2. DISCUSSION

The essential characters of concrete are durability and strength. Herein, the durability issue depends on reinforcement of material, which generally is exposed to corrosion due to some chemical and electrochemical reactions. They usually depend on the electrical conductivity of concrete and steel surface and concentration of oxygen in the pore water near the reinforcement [13]. However, the corrosion is not completely avoided. The main reasons are carbonation of concrete, and the presence of chloride ions leading to the formation of an oxide layer. This oxide layer [corrosion] is formed due to reduction of alkalinity in the concrete. The corrosion layer absorbs water or moister, which leads to the increase of volume. Therefore, the concrete cracking takes place. The service life of corroded reinforced concrete structure is comprised of two distinct phases: the first phase is corrosion initiation, and the second phase is the propagation of corrosion [14]. The first phase is related to transmission and accumulation of corrosive media. These are chloride ion and carbon dioxide; and the second one starts with corrosion of steel and ends with the failure of RC structure. The carbonation is a commonly occurring phenomenon in urban area due to the fast growth of CO_2 emission by industries. Carbonation is a chemical process that weakens the protection of concrete to the reinforcement of steel. Deterioration process from the carbonation, corrosion initiation, and accumulation of rust to cover cracking is due to uncertainties from environmental, structural, and material properties. The rate of corrosion determines the structural performance and prediction of the residual service life of RC structure. The presence of chloride ion in concrete causes more rusting than carbonation [14, 15]. If the chloride concentration around rebar exceeds the certain limit, it is referred as chloride corrosion concentration limit [CCCL]. Then the de-passivation of steel occurs, and steel is susceptible to corrosion. The corrosion rate is increased above this limit in the embedded system. The research and practices are always trying to reduce the rate of chloride ingress into the concrete with reducing the permeability of concrete and increasing the concrete length, *i.e.,* increasing concrete cover. Many, number of variables are responsible for the chloride threshold or concentration limits such as the chemical composition of cement, steel composition and temperature. Several authors said that voids on the steel surface led to the durability related problems in RC structure. The initiation of corrosion starts at the voids and interface between concrete and steel. Scientists found that chloride threshold increases as voids drop to below 2% of the interface surface. The determination of most appropriate method to find the chloride ion threshold is controversial in nature [16].

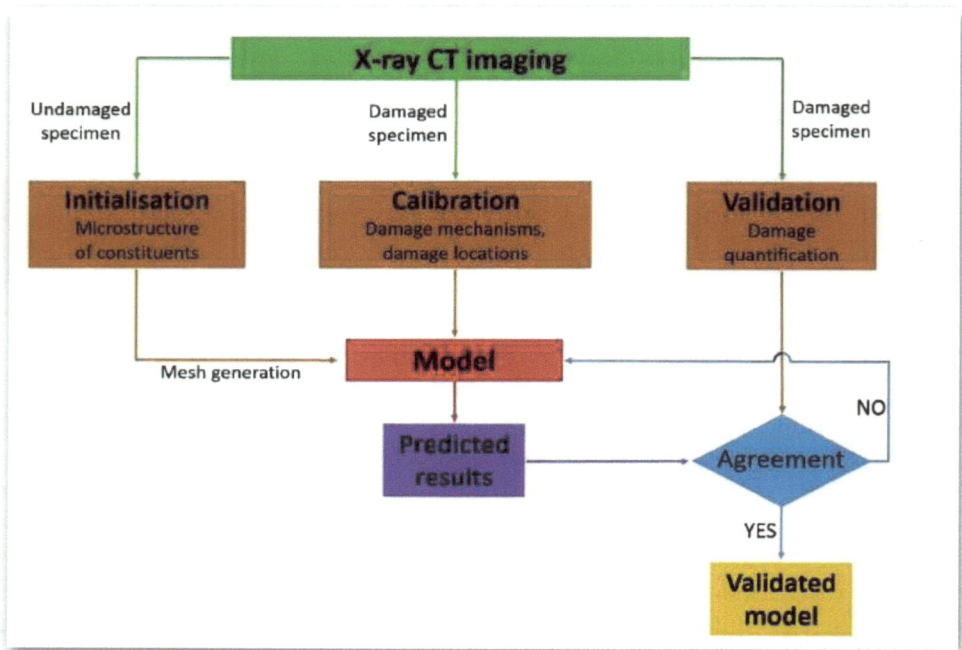

Fig. (1). The schematic representation of X-ray computed tomography (Fig. **12** of Ref [12]).

The concrete is being used across the world in large scale because of relatively low cost, easy production, and compensative strength. But the drawback is less tensile strength which is improved by the steel frame incorporation into the concrete. The steel gets less corrosion in the fresh concrete in a passive state due to the presence of high pH (12.6 to 13.5) of the pore solution. However, the ingress of carbon dioxide into alkaline components of the pore solution to activate the steel reinforcement leads to the corrosion without changing the pH value. The corrosion mitigation can be reduced by coating of epoxy layer on the steel surface [17]. The steel rebar overlapped by the passive layer is damaged by alkaline concentration. The passive layer is damaged by carbonation or chloride ion ingress, easily the steel frame can be prone to the initiation of corrosion. The steel rebar pull-out strength may be damaged by the excess temperature during summer. In America solar heat radiation leads rebar temperature to reach 80°C. Besides, the global warming also leads rising in temperature of the steel in the areas like tropical and sub-tropical [18]. In the new trend, concrete called self-compacting concrete (SCC) is more advancing the construction field than the conventional vibrated concrete. The SCC designed excellent deformability in heavy congested reinforced sections and reaches to the nook and corner of the framework without vibration for compaction. This SCC exhibits advantages because of use of secondary cementitious in place of Portland cement and reduces the acoustic noise level. Furthermore, the special characteristics of SCC are surface quality, strength, and durability [19]. The macro-cell corrosion is the most pronounced in the non-uniform deprivation of the steel which initiates at the anodic side, and the carbonation leads to micro-cell corrosion. In the carbonation, the induced macro-cell corrosion is preferred to prepare non-uniform corrosion cells. The three general approaches are used in induced macro-cell corrosion, such as 1) finite element method (FEM); 2) boundary element method (BEM); and 3) resister networks and transmission line method (RNTLM) are the most appropriate and empirical method among all methods [20].

The cover cracking time in the reinforcement structure is an important study which is occurred due to corrosion in the steel frames. Once the corrosion process is initiated, the products are formed due to electrochemical reaction, and the products can fill the porous zones between the steel and the concrete. If the zone is filled, the more space is required leading displace the concrete around the steel. That displacement creates the internal pressure on the concrete and develops the tangential stress in the cover. This leads to the cracking and opening of cover concrete. The occurrence of the first crack on the outer surface of the cover is the signal for the repair or reconstruction of the reinforced concrete. The cover cracking time can be calculated by many experimental, empirical, and numerical models. The time of corrosion is also an important factor which helps to estimate the RC structure life or durability. The corrosion initiation and speed are decided by the concrete cement ratio, the thickness of the concrete cover, surface chloride concentration, chloride diffusion coefficient, critical chloride concentration, external load, and external environment [21]. The corrosion process associated with steel comprising four complementary processes: at the anodic site, oxidation takes place, at the cathodic site, reduction of hydrogen or oxygen can take place, the moment of electrons through reinforcement and ions transportation through concrete will be taken place. If both cathodic and anodic reactions take place randomly at the neighbouring sites of same metal leads to the micro-cell corrosion. When the cathodic site and anodic sites are spatially separated, macro-cell corrosion takes place [22]. The corrosion, internal defects and mechanical properties in the reinforcement concrete can be measured by the ultrasonic tests by sending the ultrasonic waves to the structures. The elastic waves carry information containing propagation time or speed, amplitude, and frequency. These elastic waves can undergo reflection, scattering, diffraction, absorption, and wave distortion while passing through concrete. On analyzing the waves, we can glean mechanical properties and defects in the structures [23].

2.1. Impact of Bacteria on the Reinforcement Concrete

The cracking is a major problem in the reinforcement concrete due to various reasons such as shrinkage, electrochemical reactions, tensile loading, differential settlement, and thermal gradient. If there is no appropriate repair of cracks, it leads to the damage on the durability of concrete. Hence, there are many methods to repair the expansion of the crack. The traditional repairing methods are time-consuming process and have their own limitations. Of late, the microbial calcite precipitation is mixed with concrete to repair the cracks and enhance its property and reduces the corrosion in the rebar [24]. The bacterial nanocellulose fiber scan is incorporated as reinforcement in the concrete which can prevent the growth and cracking of concrete. The cellulose is prepared by organic compounds obtained by the plants, animals, and bacteria. This cellulose is renewable in nature avoiding the plastic shrinkage and increasing the strength of the concrete due to the high specific surface area of nanocellulose. Of late, in the engineering utilization, bacterial cellulose is preferred over the plant and animal cellulose. Because of the high crystalline index, high polymerization degree and purity will be occurred in the environmental point of view. The chemical formula of the bacterial cellulose is given by $(C_6H_{10}O_5)_n$ and the bacterial nanocellulose is prepared by Gluconacetobacterxylinus microorganism directly and indirectly. The self-crack healing of concrete is in advance by the bacteria agents due to the incorporation of minerals precipitating by bacterial spores with nutrients in the concrete mixture. The spores become activate when contact with water starts metabolism and precipitating calcium carbonate. The calcium carbonate is the byproduct of bacteria which can fill the cement, cracking certain degree and thereby reseal the concrete matrix [25]. There are two types of self-healing cracks in the concrete such as autogenous healing and autonomous healing. If the cracks are produced, some of them are closed because of autogenous healing. In the autogenous healing, two processes can take place like formation of calcium carbonate crystals by the calcium ions in water and continues hydration of anhydrous cement particles if the cracks are exposed to humid conditions. Thus, the calcium carbonate is formed to heal the cracks in self-healing concrete. Anyway, the autogenous healing is applicable to the crack width range from 0.01 to 0.1 mm during sufficient relevant humidity support. Hence, to overcome the drawbacks, it is replaced by autonomous healing, which can heal the cracks of greater than 0.2 mm width. The autonomous microbial self-healing is first introduced by Jonkers and Schlangen [26].

2.2. The Impact of Heat and Temperature on the Mechanical Properties of Steel

There are many types of reinforcement concretes to meet the requirements of engineering constructions for special purposes. Such reinforcement concretes are Textile Reinforcement Concrete (TRC) and Fibre Reinforced Concrete (FRC). They have their own advantages and disadvantages over a range of temperature and heat. The TRC composites are mortars along with a blend of continuous layers with equal spacing. The very few studies revealed the behavior of the TRC in the presence of fire. The temperature points such as 75°C, 150°C, and 200°C give rise to the positive properties. Thus, the mechanical property of increasing stress in the pre cracking zone is attributed. But at the elevated temperature of 400°C, it decreases the tensile strength and in between 600 to 100°C, the brittleness is failed [27]. FRC is the recent technology of withstanding reinforcement material as it is being prone to fire accidents. The withstanding capacity of reinforced concrete depends on the type of fiber incorporated such as steel fiber, glass fiber, polymer fiber, basalt fiber and natural plant fiber. On the other hand, the incorporation of replacement materials improves the performance of reinforcement at higher temperatures like silica fumes, fly ash, ground granulated blast furnace slag, metakaolin and slag powder. These incorporated materials develop denser microstructures and improvement of strength at room temperature. But on heating, the internal development vapor is difficult to estimate and it leads to spalling damage. At higher temperature (180°C to 300°C), some fibers with poor thermal stability may melt and fail the structure with the formation of fine channels inside the concrete. This acts as an outlet for the water vapors can reduce the internal pressure and heat resistant [28]. To scale down the environmental pollution, the recycled concrete is used in the place of natural concrete. The recycled concrete is made by waste concrete with the help of technology and used as road filler and buildings. This technology is conductive because it saves resources and reduces environmental degradation. But the recycled concrete is weaker than the natural concrete because of many numbers of interfaces transition zones. These interface transition zones lead to weak mechanical properties in the recycled concrete. To improve the mechanical properties of recycled concrete, there are two approaches such as 1) addition of ultra-fine mineral materials and 2) addition of fibers. In both the approaches, the modification of interface transition zones takes place. So that, it gains improved mechanical properties. In the first approach, the minerals added are ultra-fine fly ash, silicon powder and nanomaterials. In the second approach, the fibers are added *via* their bridging effect [29]. The concrete-filled steel tube (CFST) columns are used in the modern high-altitude building construction columns. These CFSTs can exhibit the high strength, ductility, and high energy absorption capacity. It even reduces the time factor as we used steel tube instead of a framework. In this work, the CFST concrete and steel tube jointly can tolerate mechanical load. The recycling scrap tires can be added into this CFST to improve the acoustics and energy dissipation, lightweight, higher impact resistance, proper crack distribution, and less cost. Any way the rubberized concrete reduces average strength. Hence, we can use this concrete for the road laying, nailing concrete and wall panels [30].

The reactive powder concrete [RPC] is promising engineering material in the construction field due to its high strength, excellent durability, and mechanical properties. This RPC has attained dense homogeneous microstructures by pozzolanic material, cement, and very fine sand (0.6 mm) with very low water/cement ratio. At the higher temperature, this RPC is prone to explosive spalling caused by vapor pressure inside the pores and thermal stress produced by temperature gradient. The vapor pressure causes the stress increased. If stress is greater than the tensile strength of RPC, the spalling can occur. So, to mitigate this spalling, the polypropylene fiber (PPF) is added into concrete [31]. In the nuclear reactors, the shielding plays a very important role to stop the radioactive radiation into the atmosphere or ecosystem. The shielding must withstand higher temperature in its service life as the thermonuclear reactions take place inside the reactor. But due to the attenuation of gamma rays, neutrons, and heavy heat flow from a hot place to cold place led to hike in temperature of concrete up to 250°C. The thick prestressed concrete shield attains up to 400°C in the reactor. As a result of the hike in temperature, the concrete structure mechanical properties will be declined. The attenuation coefficient of concrete for gamma rays and neutrons is decreased with increase in temperature. The reduction in the strength is due to the loss of physically and chemically bonded water from hydration products of cement. Then the evaporation of free water from concrete pores and the formation of thermal cracks will be happened [32].

2.3. Impact of Temperature on The Corrosion of Steel

The temperature is one of the main parameters in the process of corrosion of steel or any other alloy. In the mild carbon steel, the carbon dioxide is one of the components which can pave to the corrosion of steel. Mild carbon steel is used in the oil and gas industry. At the low pH values, the temperature is raised, and it accelerates the mass transport in bulk solution and electrochemical reactions. So that, the corrosion rate is increased. The maximum corrosion rate is attained between the temperatures 60 to 80°C. But, below 60°C the protective layer is absent, and the products of corrosion are porous in nature and no adherent to steel corrosion. Above 60°C $FeCO_3$ layer becomes dense and stick to the surface with the creation of a barrier between the metal and electrolyte. So that, it reduces the corrosion rate [33]. To reduce the corrosion problem in the oil and gas industry, the Duplex Stainless Steel (DSS) is used in the place of carbonated steel which exhibits high mechanical properties. This steel has microstructures of equal distributed ferrite and austenite parts with the absence of deleterious phases (intermetallic nitrides and carbides) [34]. The oil and gas sectors are using many technologies to increase production, and one of them is acid treatment. The utilization of acids provokes to dissolve minerals and foreign materials like drilling mud. In this process, the pumping of concentrated acids solutions to the wells takes place to increase permeability and porosity. However, during the pumping of acid solutions through the pipeline, the corrosion initiation takes place. But inhibitors are used to mitigate the corrosion [35]. Due to the over-extraction of oil and natural gas, the deposits are identified in the deeper and deeper. If we go to the deeper deposits, the common gases like CO_2 and H_2S are found with oil and natural gas. Because of this, the corrosion takes place in the pumping steel pipes. This corrosion is also caused by the temperature, liquid velocity, and chemical composition of the steel. The pitting corrosion can be increased with increasing the temperature and pressure of CO_2 gas. The H_2S gas can be dissolved in water, oil, and alcohol, which is very dangerous to human beings during mining and leads to the corrosion of steel [36]. The CO_2 is greenhouse gas can influence on the environment and threat to the natural ecology. The CO_2 can be transported through steel pipes for various purposes. In this course, the carbon steel gets corrosion. The corrosion rate is increased with increasing temperature and decreases the rate of corrosion on decreasing the temperature [37].

CONCLUSION

Sustainable development is more important than ordinary development. Today's world research on industrial materials is going on in the eco-friendly direction.

There are numerous methods to scale down the corrosion of reinforced concrete and improve the strength of the concrete. The durability of reinforced concrete against the corrosion is pertaining to the physical ability of cover in protecting the reinforcing against ingress of aggressive agents and maintenance of chemical alkaline medium with its pore solution. The innovative methods are in practice to mitigate the cracking of concrete such as bacteria and fibers. The bacteria and fibers can improve the lifetime of concrete by virtue of biochemical reactions in the concrete. The steel is the largest engineering material which is used in the construction department. It is a better alloy compared to other alloys in all physical and chemical properties. However, the complete mitigation of corrosion in the reinforcement concrete is a difficult task due to many external and internal factors of the environment.

CONSENT FOR PUBLICATION

Not applicable.

CONFLICT OF INTEREST

The author declares no conflict of interest, financial or otherwise.

ACKNOWLEDGEMENTS

Declared none.

REFERENCES

[1]   Rossi, E.; Polder, R.; Copuroglu, O.; Nijland, T.; Šavija, B. The influence of defects at the steel/concrete interface for chloride-induced pitting corrosion of naturally-deteriorated 20-years-old specimens studied through X-ray Computed Tomography. *Constr. Build. Mater.,* **2020**, *235*, 1-11.
[http://dx.doi.org/10.1016/j.conbuildmat.2019.117474]

[2]   Hong, S.; Shi, G.; Zheng, F.; Liu, M.; Hou, D.; Dong, B. Characterization of the corrosion profiles of reinforcement with different impressed current densities by X-ray micro-computed tomography. *Cement Concr. Compos.,* **2020**, *109*, 1-9.
[http://dx.doi.org/10.1016/j.cemconcomp.2020.103583]

[3]   Tang, K. Corrosion of discontinuous reinforcement in concrete subject to railway stray alternating current *Cement and Concrete Composites,* **2020**, 1-57.
[PMID: 103552]

[4]   Kamde, D.K.; Pillai, R.G. Effect of surface preparation on corrosion of steel rebars coated with cement-polymer-composites (CPC) and embedded in concrete. *Constr. Build. Mater.,* **2020**, *237*, 1-12.
[http://dx.doi.org/10.1016/j.conbuildmat.2019.117616]

[5]   Chousidis, N.; Ioannou, I.; Rakanta, E.; Koutsodontis, C.; Batis, G. Effect of fly ash chemical composition on the reinforcement corrosion, thermal diffusion and strength of blended cement concretes. *Constr. Build. Mater.,* **2016**, *126*, 86-97.
[http://dx.doi.org/10.1016/j.conbuildmat.2016.09.024]

[6]   Ibrahim, M.A.; Sharkawi, A.E-D.M.; El-Attar, M.M.; Hodhod, O.A. Assessing the corrosion performance for concrete mixtures made of blended cements. *Constr. Build. Mater.,* **2018**, *168*, 21-30.

[http://dx.doi.org/10.1016/j.conbuildmat.2018.02.089]

[7]   Eskandari-Naddaf, H.; Ziaei-Nia, A. Simultaneous effect of nano and micro silica on corrosion behaviour of reinforcement in concrete containing cement strength grade of C-525. *Procedia Manufacturing.,* **2018**, *22*, 399-405.
        [http://dx.doi.org/10.1016/j.promfg.2018.03.062]

[8]   Niu, D.; Zhang, L.; Fu, Q.; Wen, B.; Luo, D. Critical conditions and life prediction of reinforcement corrosion in coral aggregate concrete. *Constr. Build. Mater.,* **2020**, *238*, 1-11.
        [http://dx.doi.org/10.1016/j.conbuildmat.2019.117685]

[9]   Yu, H.; Da, B.; Ma, H.; Dou, X.; Wu, Z. Service life prediction of coral aggregate concrete structure under island reef environment. *Constr. Build. Mater.,* **2020**, *246*, 1-17.
        [http://dx.doi.org/10.1016/j.conbuildmat.2020.118390]

[10]   Zhou, W.; Feng, P.; Lin, H. Constitutive relations of coral aggregate concrete under uniaxial and triaxial compression *Construction and Building Materials,* **2020**, *251*, 1-14.
        [PMID: 118957]

[11]   Wang, P.; Qiao, H.; Zhang, Y.; Li, Y.; Chen, K.; Feng, Q. Three-dimensional characteristics of steel corrosion and corrosion-induced cracks in magnesium oxychloride cement concrete monitored by X-ray computed tomography. *Constr. Build. Mater.,* **2020**, *246*, 1-13.
        [http://dx.doi.org/10.1016/j.conbuildmat.2020.118504]

[12]   Garcea, S.C.; Wang, Y.; Withers, P.J. X-ray computed tomography of polymer composites. *Compos. Sci. Technol.,* **2017**, *156*, 305-319.
        [http://dx.doi.org/10.1016/j.compscitech.2017.10.023]

[13]   Sola, E.; Ožbolt, J.; Balabanić, G.; Mir, Z.M. Experimental and numerical study of accelerated corrosion of steel reinforcement in concrete: Transport of corrosion products. *Cement Concr. Res.,* **2019**, *120*, 119-131.
        [http://dx.doi.org/10.1016/j.cemconres.2019.03.018]

[14]   Xu, F.; Xiao, Y.; Wang, S.; Li, W.; Liu, W.; Du, D. Numerical model for corrosion rate of steel reinforcement in cracked reinforced concrete structure. *Constr. Build. Mater.,* **2018**, *180*, 55-67.
        [http://dx.doi.org/10.1016/j.conbuildmat.2018.05.215]

[15]   Sun, B.; Xiao, R.; Ruan, W.; Wang, P. Corrosion-induced cracking fragility of RC bridge with improved concrete carbonation and steel reinforcement corrosion models. *Eng. Struct.,* **2020**, *208*, 1-13.
        [http://dx.doi.org/10.1016/j.engstruct.2020.110313]

[16]   Kenny, A.; Katz, A. Steel-concrete interface influence on chloride threshold for corrosion – Empirical reinforcement to theory. *Constr. Build. Mater.,* **2020**, *244*, 1-13.
        [http://dx.doi.org/10.1016/j.conbuildmat.2020.118376]

[17]   Pokorný, P.; Tej, P.; Kouřil, M. Evaluation of the impact of corrosion of hot-dip galvanized reinforcement on bond strength with concrete – A review. *Constr. Build. Mater.,* **2017**, *132*, 271-289.
        [http://dx.doi.org/10.1016/j.conbuildmat.2016.11.096]

[18]   Chen, L.; Su, R.K.L. Effect of high rebar temperature during casting on corrosion in carbonated concrete. *Constr. Build. Mater.,* **2020**, *249*, 1-8.
        [http://dx.doi.org/10.1016/j.conbuildmat.2020.118718]

[19]   Jain, S.; Pradhan, B. Fresh, mechanical, and corrosion performance of self-compacting concrete in the presence of chloride ions. *Constr. Build. Mater.,* **2020**, *247*, 118517.
        [http://dx.doi.org/10.1016/j.conbuildmat.2020.118517]

[20]   Yu, Y.; Gao, W.; Castel, A.; Liu, A.; Feng, Y.; Chen, X.; Mukherjee, A. Modelling steel corrosion under concrete non-uniformity and structural defects. *Cement Concr. Res.,* **2020**, *135*, 1-16.
        [http://dx.doi.org/10.1016/j.cemconres.2020.106109]

[21]   Gao, Y.; Zheng, Y.; Zhang, J.; Wang, J.; Zhou, X.; Zhang, Y. Randomness of critical chloride

concentration of reinforcement corrosion in reinforced concrete flexural members in a tidal environment. *Ocean Eng.,* **2019**, *172*, 330-341.
[http://dx.doi.org/10.1016/j.oceaneng.2018.11.038]

[22]   Andres, B.R.; Karla, H.B.; Weerdta, K.; Geikera, M.; Geikera, M. Macrocell corrosion in carbonated Portland and Portland-fly ash concrete - Contribution and mechanism. *Cem. Concr. Res.,* **2019**, *116*, 273-283.

[23]   Xu, Y.; Jin, R. Measurement of reinforcement corrosion in concrete adopting ultrasonic tests and artificial neural network. *Constr. Build. Mater.,* **2018**, *177*, 125-133.
[http://dx.doi.org/10.1016/j.conbuildmat.2018.05.124]

[24]   Jafarnia, M.S.; Saryazdi, M.K.; Moshtaghioun, S. Use of bacteria for repairing cracks and improving properties of concretecontaining limestone powder and natural zeolite. *Constr. Build. Mater.,* **2020**, *242*, 1-12.
[http://dx.doi.org/10.1016/j.conbuildmat.2020.118059]

[25]   González, Á.; Parraguez, A.; Corvalán, L.; Correa, N.; Castro, J.; Stuckrath, C.; González, M. Evaluation of Portland and Pozzolanic cement on the self-healing ofmortars with calcium lactate and bacteria. *Constr. Build. Mater.,* **2020**, *257*, 1-11.
[http://dx.doi.org/10.1016/j.conbuildmat.2020.119558]

[26]   Rauf, M.; Khaliq, W.; Khushnood, R. A. Comparative performance of different bacteria immobilized in natural fibers for self-healing in concrete *Construction and Building Materials,* **2020**, 258-1-13.
[PMID: 119578]

[27]   Saidi. M., X.H. Vu, E. Ferrier, Experimental and analytical analysis of the effect of water content on the thermomechanical behaviour of glass textile reinforced concrete at elevated temperatures. *Cement Concr. Compos.,* **2020**, 1-38, 130690.

[28]   Wu. H.,Lin. H.,Zhou. A. A review of mechanical properties off ibrere inforced concrete at elevated temperatures. *CementandConcreteResearch.,* **2020**, *135*, 1-21.

[29]   Wang, Y.; Li, S. Peter Hughes, Fan Y. Mechanical properties and microstructure of basalt fibre and nano-silicareinforced recycled concrete after exposure to elevated temperatures. *Constr. Build. Mater.,* **2020**, *247*, 1-41.

[30]   Karimi, A.; Mahdi, N. Axial compressiveper form an ceo f steel tube columns filled with steel fiber reinforced high strength concrete containing tireaggre gate after exposureto high temperatures. *Eng. Struct.,* **2020**, *219*, 1-18.

[31]   Mazin, A.R.; Alyaa, A.A.; Hussein, M.; Hamadac, B.T. Microstructure and structural analysis of polypropylene fibre reinforced reactive powder concrete beams exposed to elevated temperature. *J. Build. Eng.,* **2020**, *29*, 1-8.

[32]   Demir. I., M. Gümüs, H.S. Gökçe, Gamma ray and neutron shielding characteristics of polypropylene fiber-reinforced heavyweight concrete exposed to high temperatures. *Constr. Build. Mater.,* **2020**, *257*, 1-9.

[33]   Rizzo, R.; Baier, S.; Rogowska, M.; Ambat, R. An electrochemical and X-ray computed tomography investigation of the effect of temperature on CO_2 corrosion of 1cr carbon steel. *Corros. Sci.,* **2020**, *166*, 1-38.
[http://dx.doi.org/10.1016/j.corsci.2020.108471]

[34]   Tavares, S.S.M.; Batista, R.T.; Landim, R.V.; Velasco, J.A.; Senna, L. Investigation of the effect of low temperature aging on the mechanical properties and susceptibility to sulfide stress corrosion cracking of 22%Cr duplex stainless steel. *Eng. Fail. Anal.,* **2020**, *113*, 1-20.
[http://dx.doi.org/10.1016/j.engfailanal.2020.104553]

[35]   Solomon, M.M.; Umoren, S.A.; Quraishi, M.A.; Tripathi, D.; Abai, E.J. Effect of akyl chain length, flow, and temperature on the corrosion inhibition of carbon steel in a simulated acidizing environment by an imidazoline-based inhibitor. *J. Petrol. Sci. Eng.,* **2020**, *187*, 1-39.

[http://dx.doi.org/10.1016/j.petrol.2019.106801]

[36]   36] Asadian, M., Saɔzi, M., &Anijdan, S. H. M. The effect of temperature, CO2, H2S gases and the resultant iron carbonate and iron sulfide compounds on the sour corrosion behaviour of ASTM A-106 steel for pipeline transportation. *Int. J. Press. Vessels Piping,* **2019**, *171*, 184-193. [http://dx.doi.org/10.1016/j.ijpvp.2019.02.019]

[37]   McCarthy, M.J.; Tittle, P.A.J.; Dhir, R.K. Corrosion of reinforcement in concrete containing wet-stored fly ash. *Cement Concr. Compos.,* **2018**, *73*, 60-69.

CHAPTER 7

Environmental Cracking of High-Strength Aluminum Alloys

B. Venkata Shiva Reddy[1,2], N. Suresh Kumar[3], K. Chandra Babu Naidu[1,*], M. Balaraju[4] and T. Anil Babu[1]

[1] *Department of Physics, GITAM Deemed to be University, Bangalore-562163, Karnataka, India*

[2] *Department of Physics, The National College, Bagepalli-561207, Karnataka, India*

[3] *Department of Physics, JNTU College of Engineering Anantapur, Anantapuramu-515002, A.P., India*

[4] *Department of Chemistry, PSC & KVSC Govt. Degree College, Nandyal, Kurnool-518502, A.P., India*

Abstract: The aluminium alloy is the second-largest alloy being used in the world next to steel. Aluminium exhibits good mechanical strength, resistant to corrosion and lightweight. But due to some environmental variations or conditions, the alloy was found to be prone to cracking. The microstructure cracking is due to many factors like corrosion, mechanical stress, thermal stress, and bacterial adherence. The Sulfate Reducing Bacteria (SRB) is the most active bacterium, which causes rusting. The chloride ions present around the aluminium alloy led to corrosion and the failure of microstructures in the alloy.

Keywords: 3D printing, Alloys, Corrosion, Cracking, Microstructures.

1. INTRODUCTION

An alloy is a mixture of two or more metals mixed in a required proportion. This gives special and improved properties than single metal. Such properties are strength, ductility, malleability, conductivity, resistivity, *etc.*, used in industrial sectors. Aluminium alloy is the second-largest industrial material next to steel and has many advantages in engineering materials. In recent context, aluminium-based alloy materials are being manufactured by 3D printing or additive manufacturing [AM] rather than the traditional manufacturing process. 3D printing is preferred due to unlimited freedom design, economical, less time consuming, no wastage of substance, *etc.* [1]. In additive manufacturing, varieties

* **Corresponding author K. Chandra Babu Naidu:** Department of Physics, GITAM Deemed to be University, Bangalore-562163, Karnataka, India; Tel: +91-9398426009; E-mail: chandrababu954@gmail.com

N. Suresh Kumar, P. Banerjee, H. Manjunatha and K. Chandra Babu Naidu (Eds.)

of methods are being used to print the materials, but in aluminium alloy printing, the preferable method is selective laser melting [SLM]. Anyway, there are common problems in the SLM such as cracking, balling and porosity [2]. Fatigue is one of the most prominent problems associated with structures, as the object is subjected to cyclic load. It is well recorded in many studies that the crack nucleation does not contribute particularly to the total fatigue of the component. Moreover, most of the fatigue life will be in short crack growth rather than long crack. The short fatigue cracks are in the order of 10 μm to 1 mm dimensions. Short cracks are classified broadly into; chemically short cracks, physically short crack, microstructurally short crack, and mechanically short crack [3]. The skin of the aircraft is made up of aluminium alloy, which accounts for 50% of its total weight. The micro-cracks will be taken place due to various external factors like heavy load or stress, temperature, and pressure [4]. Among many aluminium alloys, the cast aluminium alloy (A356Al Alloy) is widely used in automobiles like pistons, engine parts and cylinder heads. In automobile heat engines, cast aluminium can withstand high temperatures, and it possesses high strength to weight ratio and wear resistance.

However, α-aluminium dendrites and interdendritic irregular in Al-Si eutectic (easily melting) regions lead to microstructural defects in the system. To overcome these problems, many approaches are available such as modification of microstructures and friction stir processing [5]. The 7XXX series of aluminium alloys are widely used in military and aerospace sectors. They possess low density, supreme mechanical properties, and outstanding machinability. In 7A99 Al alloy, high Zn and low Cu mixtures were developed with high mechanical strength and low corrosion. But the artificial ageing of the aluminium alloy leads to structural, physical, and chemical changes such as microstructures, mechanical properties, stress, and corrosion cracking. To improve the mechanical properties, the intense grain refinement, and achieving the minimum sub micrometric scale (>1 μm) through severe plasticity [6, 9] are adopted. However, the thin oxide layer is formed around the aluminium alloys, preventing the destructive corrosion process. The 5XXX (Al-Mg) series are more corrosive resistant than 4XXX (Al-Si) and 6XXX (Al-Si-Mg) grades. Anyway, the presence of excess Mg leads to corrosion. Hence, a new grade AA5052 with 2 to 3% of Mg is allowed, which can curb the corrosion. In the AA5052 grade, it is very hard to form a continuous intermetallic phase located at grain boundaries [7].

During the laser welding process in Al-Mg composition alloy, the Mg will be evaporated. This leads to the weld material, and evaporation of magnesium metal can affect the solidification path of the alloy [8]. The equal channel angular pressing (ECAP) can improve the mechanical properties of aluminium alloys like Al-Mg, Al-Cu, Al-Si, *etc.*, by means of intense grain refinement. The halide atmosphere around the aluminium alloy leads to pitting corrosion, and the reduction of grain size contributes to the minimization of corrosion. The minimization of grain boundary size leads to the increase in grain boundary density and breakdown of secondary phase particles below the critical size [9]. The minute changes in the service environment account for the major damages in alloyed structures. In 1985, Japan airlines flight 123 prone to fatal with 520 lives due to long-term corrosion fatigues [10]. The temperature, relative air humidity and corrosive media play a crucial role in the aluminium alloy properties. It is found that at lower temperatures below -20°C, all alloys can exhibit longer fatigue lives and shorter fatigue crack propagation rate (FCGRs). But at room temperature, the alloy exhibits higher strength and fatigue lives decrease. If the temperature exceeds 350°C, it decreases the strength and increases the oxidation rate. With increasing temperature, relative air humidity decays the fatigue properties of the alloy. This leads to the acceleration of FCGRs of aluminium alloy. In corrosive media like NaCl fog, acid solution and alkaline solution, aluminium alloys are quite sensitive. Thus, the initiation and propagation of micro-cracks take place. In the Na-Cl solution, Cl ions cause severe pitting on the surface [10, 11].

Solidification is one of the causes to get of fracture in alloys, and it is called solidification cracking (SC). Between the solidified weld metal and welding pool, there exists a semi-solid region, that is a mushy zone. Solidification cracking takes place at the end of mushy zone, where some residual liquids can exist in the form of film in a liquid state. The initiation of SC takes place at the terminal solidification stage due to the reasons such as shrinkage, thermal strain, hindered contraction and lack of liquid feeding [12]. However, aluminium alloys are easily prone to cracking during solidification because solid density is greater than liquid density. Further, another problem; the micro-voids are also formed in the brittle secondary phase due to the initiation of microcracks. Thus, lead to a failure in the alloys [13]. During the welding alloys, the microstructural guarantee is important and maintained by friction stir welding (FSW). This exhibits excellent weld efficiency and avoids high heat entry during manufacture. The advanced welding system called Bobbin tool friction stir welding (BT-FSW) is used. The BT-FSW requires back support and particularly suitable in closed structures like pipes and tanks. Of late, semi-stationary shoulder variant (SSuBT-FSW) is the working concept in alloys like Al-Cu-Li [14]. However, the fatigue properties are anisotropic in nature and to scale down the cracking susceptibility, filter metals along with non-matching composites are often essential to adjust the composition of the weld metal in the fusion zone [15]. The ageing factor of alloy also plays a very important role in fatigue and cracking, such as peak aged, over aged, retrogression and re-ageing. The re-ageing alloy exhibits the highest resistance to the fatigue crack initiation, and in the peak, the aged crack propagation and growth rate are maximum and smallest for re-ageing [16].

Based on the metal theory and atomic diffusion theory, varieties of advanced laminated composites are prepared by some techniques such as 1) accumulative roll bonding 2) hot extrusion 3) composite welding and 4) explosion cladding. Of late, Equal angular extrusion (ECAP) and High-pressure torsion (HPT) are extensively used to produce laminated alloy composite sheets. With the relevant extrusion temperature and speed, we can prepare alloy sheets along with excellent mechanical and high precision. However, in the multilayer alloy sheets at a lower temperature, during hot extrusion of the compound and in phase bending torsion loading, the fatigue cracks will be initiated [17]. The multiaxiality will be raised due to many factors such as multiaxial loading, complex geometry, residual stress, crack orientation, *etc*. However, mode-I and mode-II loading conditions are mixed responses to the fatigue loading [18].

2. DISCUSSION

2.1. The Role and Mitigation of Corrosion

The aluminium alloy has several advantages by virtue of its low density, high mechanical strength, ductility and broad applications in aerospace, automobiles, and oil industry. Even though aluminium alloy is reluctant to corrosion by virtue of thin oxide layer formation on the surface, during some conditions, it suffers from corrosion leading to micro cracking. Such conditions are acid and alkaline solutions, defects in the oxide layer and doping of other elements like Fe, Mn, S *etc*., for example, pitting corrosion can exist in the NaCl solution when the alloy is immersed [19]. Due to the presence of Cl ions in the aqueous solutions, the aluminium alloy (*e.g.* AA6061) suffers from corrosion. Organic compounds coating and surfactants can curb the corrosion, but not effectively. Hence, the rare earth elements can be incorporated into alloys. The rare earth compounds are used in this alloy to curb the rusting such as $CeCl_3$ and $Ce_2(SO_4)_3$, which blanks the chloride solution to stop the corrosion initiation. The rare earth compounds are also eco-friendly, low cost and can be easily prepared [20]. In the seawater, environment alloys are prone to be corroded due to the disparity of relative nobility in each phase. The selective phase corrosion leads to local damage, percolated into the substrate. Then, it produces a surface structure which decides mechanical strength and avoids hazardous constructions. The LASER surface quenching is an effective method to avoid corrosion initiation and modifies the microstructures in the alloy in the sea environment. When the laser light is irradiated on the surface of the alloy, it is heated above the phase transition temperature suddenly by thermal effect and suddenly, at a rapid rate it is cooled down. This sudden process is called laser quenching possessing high treatment efficiency, controllable quenching depth, and low quenching stress. In the laser quenching, heating, and cooling are extremely fast, about 1010°C/s, to fulfill the aim of single-phase [21]. The schematic representation of laser surface quenching is shown below in Fig. (**1**).

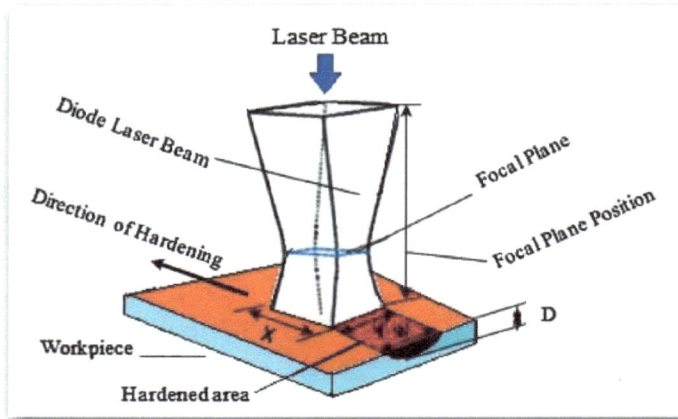

Fig. (1). Schematic representation of laser surface quenching (Fig. **1** of Ref [22]).

The aluminium alloy AA-6061-T6 is used in Material Testing Reactors (MTRs) due to its low activation, good neutron transparency and excellent mechanical properties. The coolant used in this reactor is typically nitric acid diluted in pure water to get pH between 5 and 6. The temperature of the core structure outside the fuel typically reaches between 70 and 100°C. In this context, aluminium alloy can easily be prone to corrosion. Thus, the aluminium hydroxide film is formed covering the alloy. Due to the low thermal conduction of hydroxide (2 W/m/K), the film degrades heat exchange between core structure and coolant. Finally, it leads to overheating of the core producing corrosion [23]. To avoid corrosion in this context, we can prepare efficient ultra-thin films which are able to nanoscopic molecular control, diversity, and flexible design. The formation of an organic monolayer on the passive film improves the pitting potential of the metal substrate [24]. The aluminium alloy 7075 suffers from corrosion. Hence, we can employ a new technique called the Cold Spray (CS), and it is known as cold sprayed aluminium alloy. The cold sprayed aluminium alloy 7075 is novel and produced by innovative 3D printing or additive manufacturing. In the cold spray, alloyed particles are being accelerated by a stream of supersonic gas and deposited on the substrate under some parameters such as pressure and temperature. The sprayed particles are suffered from some plastic deformation. Compare to conventional spray, the oxidation, phase transformation, and excessive thermal stress can be effectively scaled down because powder particles remain deposited in the bottom. This cold spray can protect the alloy from higher temperature corrosion, oxidation reactions and chemical reactions. Of late, an improved version of cold spray is in practice that is High-Pressure Cold Spray (HPCS). This can repair the damaged components instead of replacement of components. This HPCS can also be useful in solid construction on un-weldable base metals, and the substrate will remain unaltered [25].

In the manufacture of cars, the predominantly used alloy is 7N01 alumina alloy. This exhibits lightweight and low maintenance cost and extruded into thin walls with complex structures. Hence, they are used in corbels, bases, section beams and bearings of trains and vehicles. However, the aluminium alloy 7N01 is relatively thin, poor wear and corrosive resistant. The aluminium alloy cannot withstand for a long time and not able to maintain in severe conditions. Hence, it is inevitable to develop the surface properties with the treatment of new technology. The Micro Arc Oxidation (MAO) is a relevant approach to improve the surface properties of the 7N01 aluminium alloy. The MAO is a surface treatment and improvement technology developed from anodic oxidation. The MAO produces the micro-arc plasma discharge which envisages ceramic coating on the surface of the 7N01 aluminium alloy. Thus, it develops resistance against corrosion, wear resistance and a solid combination of matrix metal [26]. The 2A14 aluminium alloy is a mixture of Al-Cu-Mg and adding Si into this alloy develops its age-hardening response. This alloy is used in the aerospace and aircraft industry by virtue of its excellent properties such as high specific strength, workability, and weldability. No doubt, that 2A14 alloy possesses good properties and suitable to use in aerospace and aircraft industry. But further improvement of the alloy is essential to advance developments. Hence, we can employ solid solution and ageing treatment. The single stage peak ageing (T6) is a commonly used technology to improve or maximize the strength of the 2A14 aluminium alloy and reduces the stress corrosion resistance. In the two-stage peak ageing (T7X) scale of the plasticity and corrosion resistance of aluminium alloy (2A14 but strength is reduced to 4 to 8%) is increased. Hence, we employed multistage ageing treatment to improve mechanical properties and corrosion resistance of 2A14 aluminium alloy by virtue of continuous heating and cooling of alloy [27].

The onshore resources are gradually declined due to over-extraction and utilization. Hence, we should investigate sea or ocean resources in present days. Compare with the onshore environment, the sea environment is different with regards to hydrostatic pressure, a higher level of salinity and rich marine population. In marine conditions, aluminium steel can easily be prone to corrosion leading to cracking of microstructures. The passive film formed on the alloy surface is damaged and reconstruction of film is hindered by chloride ions and microorganism. The Desulfovibriosp caused severe corrosion in Q235 aluminium alloy in marine condition. To mitigate the corrosion in a marine context, we can coat organic materials such as graphene oxide-modified polyaniline and doped zinc. Even though this organic coating is done, the alloy leads to corrosion due to variation of environmental conditions. Moreover, it suffers from many defects due to long exposure marine environment. The biofilms and metabolite products of microorganisms impact negatively on the corrosion behavior of metals. Some microorganisms contain extracellular substances providing a barrier to stop corrosion. The Pseudoalteromonassp and Pseudomonas fragi K or Bacillus brevi bacteria form the aerobic film and inhibit the corrosion in the marine environment [28].

The bacterial corrosion is a big challenge to the scientific community. The sulfate-reducing bacteria (SRB) is the most corrosive bacterium which presents in the soils, pipes, and marine environment. The SRB damages not only aluminium alloy, but also iron and copper. It reduces the sulfate into sulfide with the utilization of organic matter in an aqueous solution as a carbon source. The chemical attack from hydrogen sulfide and extracellular substances secreted by SRB leads to corrosion. The SRB acts as a biocathode which obtains the electrons directly from metals with extracellular electron transfer (EET) in the presence of lack of carbon source. The presence of electron mediators and carbon starvation will promote the EET of SRB, leading to pitting corrosion [29]. The Aspergillus Niger is a class of the SRB which causes more corrosion than any bacteria living in warm and humid conditions. The bacteria can grow in the range of 6 to 47°C and active in the pH range of 1.4 to 9.8. It induces corrosion on anodized and non-anodized 6061 aluminium alloy. The metabolic activities play a very important role in bio-corrosion process by the Aspergillus Niger. These metabolites are citric acid, oxalic acid, malic acid, gluconic acid, succinic acid, and pectinases. These metabolites are the main source of corrosion of aluminium alloy steel or any metals or alloys [30].

2.2. The Role of Fatigue in Cracking of Aluminum Alloy

Fatigue failure is one of the basic problems that take place in automotive engineering parts which work under dynamic as in case of control arms of the suspension system. This failure takes place under dynamic loading conditions. They occur without warning. This failure is a sudden process and divided into four stages such as cyclic stress, crack initiation, crack propagation and fracture. The fluctuation stress cycles are the root cause of fatigue failure caused by completely reversed stress cycles, repeated stress cycles and random stress cycles.

The random stress cycles are responsible for fatigue failure in the automotive engineered parts [31]. To join the aluminium alloy sheets or plates, the common method is riveted joints. When tensile loading is applied, tensile loads are transferred through aluminium plates at fastener holes. The stress fields around the fastener holes are critical, where the stress is concentrated, and nucleation of fatigue cracks take place [32]. To improve the fatigue strength of aluminium alloy, various technologies are in effect in the engineering domain such as Shot peening (SP) and Cavitation peening (CP). The SP is most practicing technology to improve the fatigue strength in aerospace applications and improves the rotating bending fatigue life of 7075 aluminium alloy. This exhibits comprehensive residual stress on the surface. In the Cavitation peening, there is no question of shot and improves the fatigue strength with plastic deformation of the substrate through the impact of the cavitation bubble. In the CP, there is no surface roughness significantly, because it does not take part in physical contact between the solids. From this CP treatment, the silicon-manganese alloy bending fatigue strength, aluminium casting alloy strength and carbonized chrome-molybdenum steel strength are successfully improved [33].

The fatigue theory regarding multiaxial loading can be explained by two approaches, one is a classical stress/strain method, and other is damage mechanic's approach. The classical stress/strain approach is well established in the literature and depends on the empirical observations from the fatigue data consisting of two groups such as the critical plane group and stress invariant group. The damage mechanics (DM) approach is well advanced in monotonic applications and expanded to the cyclic conditions. The DM theory can be figure out by two approaches such as the micro mechanism of defect models and Continuum Damage Mechanism (CDM). The DM theory secured a special interest from the recent years regarding cyclic applications. This is independent of any cyclic counting methods of fatigue damage application rule, but it depends on the load history. Thus, this DM approach is largely applied to random and complex loads. The CDM approach reached results that are more realistic and applied. This is a theory for fatigue life predictions in adhesively bonded joints with respect to random load history. CDM approach is a powerful technique, but it depends on the presence of macro plastic strains [34]. The aluminium alloy Al6061-T6 is preferred in automotive due to its light weight. This reduces the fuel consumption and environmental pollution by reducing CO_2 emission. However, the aluminium alloy Al6061-T6 has no definite fatigue limit and possess larger scatter in fatigue crack growth rate when compared to steel. Hence, the doping is done on the Al6061-T6 alloy to improve the fatigue limit, and such dopants are Mg and Zr. Herein, mere addition of Mg leads to limit the crack propagation. But the excess Mg doping increases the grain size leading to decrease the tensile strength. So, to solve this problem, excess Mg and Zr are added into Al6061-T6 alloy, which exhibits fine grain size and tensile strength [35]. The commercial aluminium alloy 7000 series exhibit very good characteristics in many engineering applications. In commercial 7075 alloy, cracks will be nucleated at the inter-metallic junction, where the failure is propagated. For example, Fe and Si are the impurities which can form inclusion leading to cracking in the alloy. When the 7000-alloy series are exposed to corrosive and fatigue loading, automatically alloy is prone to the failure and decreases the mechanical properties. The cracks are being initiated by virtue of corrosion pits on the surface. The surface coating is effectively employed for the mitigation of fatigue in the presence of corrosion. But conversely, cracks and defects will be formed during pre-surface treatment and etching in anodization treatment. Hence, advanced coating is given to the surface by TiN, ZrN, WC/C, NieP and DLC Diamond-like carbon) PVD (Physical vapour deposition) coating to improve the fatigue strength in the corrosive medium. In the PVD process, high temperature is produced near to the solidification alloy. Therefore, it reduces the mechanical properties of the substrate. For this reason, we need post-deposition heat treatment to improve fatigue and tensile strength. Any way at low temperature conditions, the PVD and DLP give out good mechanical properties of alloys [36]. The multi-component alloys like Nickel-Aluminum Bronze (NAB) with iron and manganese as major alloying components. These are widely used in marine components as a propeller material as it possesses a cluster of favorable properties such as mechanical strength, fracture toughness and corrosive resistance. Scientists reported that selective phase corrosion and lamellar α phases in $\alpha + K_{III}$ eutectic structures are predominantly common problems in the NAB alloy. The corrosive fatigue failure of metals is classified into three stages such as crack initiation, crack propagation and final fracture [37].

2.3. Impact of Temperature on Aluminum Alloy

The heat engine components are subjected to both thermal and mechanical stress in which the thermal load occurs in an isothermal process. The components used in the heat engine exhibited a temperature gradient and made up of age-hard enable aluminium alloy. The thermal load on this alloy leads to distraction in complex microstructures. The rate of the distraction of microstructures depends on the type of alloy in which we used heating components. The process of heating and quenching of solids leads to the formation of small clusters in the matrix. The cluster growth and resulting change in material properties lead to damage of microstructure in the heat engine components. With the increase of temperature, the nucleation and growth rate are also increased due to higher diffusion rate [38]. The new generation aluminium alloy is 7085 with high strength and low quench sensitivity. It is kind of typical age hard enable alloy because it forms the precipitation during heat treatment. The different ageing leads to different microstructures to form different physical, mechanical, and thermal properties of the alloy. The good ageing temperature is 120°C for the 7085 alloys in which we get good morphology and microstructure size [39]. The 7XXX aluminium alloy exhibits excellent properties as zinc content is a major alloying element which can withstand to hot workability. The microstructure characteristics such as precipitation behavior, grain size, mechanical properties, and corrosive resistance, entirely depend on the heat treatment. During the single ageing treatment in 7XXX series aluminium alloy, they provide good mechanical properties but lags in corrosive resistance. Besides, in the double ageing treatment, corrosive resistance can be increased by reducing 10 to 15% of mechanical strength rather than that of single ageing treatment [40].

The conventional welding of aluminium alloy is difficult to process. Hence, we prefer an alternative method to overcome the difficulties. The novel method of welding is Friction Stir Welding (FSW) in which temperature plays an important role as compared to fusion weld. In the FSW welding, the dissimilar metals can be joined easily, and friction takes place between the work piece and tool shoulder interface. During friction, extensive heat is generated, which plays an important role in thermo-mechanical welding. The friction and the galling behavior of alloys depend on many factors like sliding velocity, contact pressure, the geometry of the tool and sliding contact surface of the contact. The heat generated depends on the speed of rotations of the tool, the temperature is increased if the rotation speed is varied on decreasing the speed of rotation. Thus, the strength of a weld joint depends on the friction pressure. The friction coefficient is increased with increasing temperature [41, 42]. The aluminium alloys should withstand the higher temperature alternative loading during service. Therefore, high temperature fatigue performance gains much importance in aluminium alloys. To improve the mechanical properties further, Laser shock peeing (LSP) is employed by the present researchers. The LSP increases wear and corrosive resistance along with increasing high temperature fatigue. Many researchers noticed that LSP provides better resistance to the thermal cyclic loads than conventional methods like a shot peeing [43].

CONCLUSION

The aluminium alloy is the most predominant alloy next to steel alloy used in the present world for different development and constructions. The aluminium alloy is unique in nature because it can exhibit many engineering applications in the present technology. In addition, fast track research is going on to enhance the characteristics, properties and to invent new properties. The properties of alloy can be manipulated with the addition or doping of other elements like zinc, silicon, manganese, and chromium. The different aluminium alloys are manufactured according to the requirements. The lightweight nature of this alloy showed many implications in the automobiles and aerospace domain. This lightweight nature reduces fuel consumption and helps to reduce CO_2 emission into the atmosphere. After the invention of 3D printing or additive methods, the aluminium alloy materials are being printed more accurately with decreasing manufacturing defects than conventional methods. In the conventional manufacturing methods, many defects were noticed. Hence, the present manufacturing alloys are the outcomes of 3D printing. Even though aluminium alloy exhibits many advantages over other alloys, many disadvantages will be noticed. The major problems are cracking of microstructures and rusting, which are due to external factors like change in temperature, over cyclic mechanical fatigue load and thermal pressure. Even in the presence of bacteria, the atmosphere leads to the corrosion and finally leads to the collapse of aluminium alloy frames. Hence, there is a lot of responsibility on the scientific and research community to develop risk-free aluminum alloys to the society.

CONSENT FOR PUBLICATION

Not applicable.

CONFLICT OF INTEREST

The author declares no conflict of interest, financial or otherwise.

ACKNOWLEDGEMENTS

Declared none.

REFERENCES

[1]   Aboulkhair, N.T.; Simonelli, M.; Parry, L.; Ashcroft, I.; Tuck, C.; Hague, R. PMSP Aboulkhair 3D printing of Aluminium alloys. *Prog. Mater. Sci.,* **2019**, *106*, 1-45, 100578.
        [http://dx.doi.org/10.1016/j.pmatsci.2019.100578]

[2]   Li, R.; Wang, M.; Li, Z.; Cao, P.; Yuan, T.; Zhu, H. Developing a high-strength Al-Mg-Si-Sc-Zr alloy for selective laser melting: crack-inhibiting and multiple strengthening mechanisms. *Acta Mater.,* **2020**, *193*, 83-98.
        [http://dx.doi.org/10.1016/j.actamat.2020.03.060]

[3]   Vidit, G.; Manabu, E.; Syohei, Y. Physically short and long-crack growth behavior of MIG welded Al5.8%Mg alloy. *Eng. Fract. Mech.,* **2019**, *209*, 301-316.
        [http://dx.doi.org/10.1016/j.engfracmech.2019.01.026]

[4]   Wu, L.; Wang, T.; Hu, Y.; Liu, J.; Song, M. A method for improving the crack resistance of aluminum alloy aircraft skin inspired by plant leaf. *Theor. Appl. Fract. Mech.,* **2019**, *106*, 1-20, 102444.
        [http://dx.doi.org/10.1016/j.tafmec.2019.102444]

[5]   Nelaturu, P.; Jana, S.; Mishra, R.S.; Grant, G.; Carlson, B.E. Effect of temperature on the fatigue

cracking mechanisms in A356 Al alloy. *Mater. Sci. Eng. A,* **2020**, *780*, 1-26, 139175.
[http://dx.doi.org/10.1016/j.msea.2020.139175]

[6]   Hou, Y.; Chen, L.; Li, G.; Zhao, G.; Zhang, C. Effects of artificial aging on microstructure, mechanical properties and stress corrosion cracking of a novel high strength 7A99 Al alloy. *Mater. Sci. Eng. A,* **2020**, *780*, 1-27, 139217.
[http://dx.doi.org/10 1016/j.msea.2020.139217]

[7]   Shuwei, D.; Dongting, W.; Wenyu, L.; Tao, W.; Yong, Z. Stress corrosion cracking behavior of friction stir welded 5052-H112 Al–Mg alloy. *Vacuum,* **2020**, *176*, 1-8, 109299.
[http://dx.doi.org/10.1016/j.vacuum.2020.109299]

[8]   MalekshahiBeiranvand, Z.; MalekGhaini, F.; NaffakhMoosavy, H.; Sheikhi, M.; Torkamany, M.J. Solidification cracking susceptibility in pulsed laser welding of Al–Mg alloys. *Materialia,* **2019**, *7*, 1-9.
[http://dx.doi.org/10.1016/j.mtla.2019.100417] [PMID: 100417]

[9]   DiogoPedrino, B.; Danielle Cristina, C.M.; Andrea, M.K.; Carlos Alberto, D.R.; Vitor, L.S. Microstructure, mechanical behavior and stress corrosion cracking susceptibility in ultrafine-grained Al-Cu alloy. *Mater. Sci. Eng. A,* **2020**, *773*, 1-10, 138865.
[http://dx.doi.org/10.1016/j.msea.2019.138865]

[10]   Chena, Y.Q.; Zhanga, H.; Songa, W.W.; Panc, S.P.; Liu, W.H.; Liu, X.; Liu, B.W.; Songa, Y.F.; Zhoua, W. Acceleration effect of a graphite dust environment on the fatigue crack propagation rates of Al alloy. *Int. J. Fatigue,* **2019**, *216*, 20-29.
[http://dx.doi.org/10.1016/j.ijfatigue.2019.04.033]

[11]   Thurston, K.V.S.; Gludovatz, B.; Yu, Q.; Laplanche, G.; George, E.P.; Ritchie, R.O. Temperature and load-ratio dependent fatigue-crack growth in the CrMnFeCoNi high-entropy alloy. *J. Alloys Compd.,* **2019**, *794*, 525-533.
[http://dx.doi.org/10.1016/j.jallcom.2019.04.234]

[12]   Geng, S.; Jiang, P.; Shao, X.; Mi, G.; Wu, H.; Ai, Y.; Wang, C.; Han, C.; Chen, R.; Liu, W.; Zhang, Y. Effects of back-diffusion on solidification cracking susceptibility of Al-Mg alloys during welding, A phase-field study. *Acta Mater.,* **2018**, *160*, 85-96.
[http://dx.doi.org/10.1016/j.actamat.2018.08.057]

[13]   Cui, X.; Yu, Z.; Liu, F.; Du, Z.; Bai, P. Influence of secondary phases on crack initiation and propagation during fracture process of as-cast Mg-Al-Zn-Nd alloy. *Mater. Sci. Eng. A,* **2019**, *759*, 708-714.
[http://dx.doi.org/10.1016/j.msea.2019.05.062]

[14]   Jannik, E.; Martina, M.; Martin, R.; Carsten, B.; Mikhail, Z.; Joge, F. The effect of grain boundary precipitates on stress corrosion cracking in a bobbin tool friction stir welded Al-Cu-Li alloy *Materials Letters: X, X2,* **2019**, (), 1-4.
[http://dx.doi.org/10.1016/j.mlblux.2019.100014] [PMID: 100014]

[15]   Soysal, T.; Kou, S. Effect of filler metals on solidification cracking susceptibility of Al Alloys 2024 and 6061. *J. Mater. Process. Technol.,* **2019**, *266*, 421-428.
[http://dx.doi.org/10.1016/j.jmatprotec.2018.11.022]

[16]   Yao, L.; Guofu, X.; Shichao, L.; Xiaoyan, P.; Zhimin, Y.; Li, W.; Xiaopeng, L. Effect of ageing treatment on fatigue crack growth of die forged Al5.87Zn-2.07Mg-2.42Cu alloy. *Eng. Fract. Mech.,* **2019**, *215*, 251-260.
[http://dx.doi.org/10.1016/j.engfracmech.2019.04.023]

[17]   Sheng, K.; Lu, L.; Xiang, Y.; Ma, M.; Wu, Z. Crack behavior in Mg/Al alloy thin sheet during hot compound extrusion *Journal of magnesium alloy,* **2019**, *7*, 717-724.
[http://dx.doi.org/10.1016/j.jma.2019.09.006]

[18]   Abhay, K. Singha, SiddhantDattaa, Aditi Chattopadhyaya, Jaret C. Riddickb, Asha J. Hallb, Fatigue crack initiation and propagation behavior in Al – 7075 alloy under in-phase bending-torsion loading.

Int. J. Fatigue, **2019**, *126*, 346-356.
[http://dx.doi.org/10.1016/j.ijfatigue.2019.05.024]

[19]   Zhang, K.; Yang, W.; Ge, F.; Xu, B.; Chen, Y.; Yin, X.; Liu, Y.; Zuo, H. A self-curing konjac glucomannan/CaCO$_3$ coating for corrosion protection of AA5052 aluminum alloy in NaCl solution. *Int. J. Biol. Macromol.,* **2020**, *151*, 691-701.
[http://dx.doi.org/10.1016/j.ijbiomac.2020.02.223] [PMID: 32088236]

[20]   Deyab, M.A.; El-Rehim, S.S.A.; Hassan, H.H.; Shaltot, A.M. Impact of rare earth compounds on corrosion of aluminum alloy (AA6061) in the marine water environment. *J. Alloys Compd.,* **2020**, *820*, 1-35, 153428.
[http://dx.doi.org/10.1016/j.jallcom.2019.153428]

[21]   Qin, Z.; Xia, D.; Zhang, Y.; Wu, Z.; Liu, L.; Lv, Y.; Hu, W. Microstructure modification and improving corrosion resistance of laser surface quenched nickel-aluminum bronze alloy. *Corros. Sci.,* **2020**, *174*, 1-39, 108744.
[http://dx.doi.org/10.1016/j.corsci.2020.108744]

[22]   Moradi, M.; Arabi, H.; JamshidiNasab, S.; Benyounis, K.Y. A comparative study of laser surface hardening of AISI 410 and 420 martensitic stainless steels by using diode laser. *Opt. Laser Technol.,* **2019**, *111*, 347-357.
[http://dx.doi.org/10.1016/j.optlastec.2018.10.013]

[23]   L'Haridon-Quaireau, S.; Laot, M.; Colas, K.; Kapusta, B.; Delpech, S.; Gosset, D. Effects of temperature and pH on uniform and pitting corrosion of aluminium alloy 6061-T6 and characterization of the hydroxide layers. *J. Alloys Compd.,* **2020**, *883*, 1-12, 155146.
[http://dx.doi.org/10.1016/j.jallcom.2020.155146]

[24]   Xia, D-H.; Pan, C.; Qin, Z.; Fan, B.; Song, S.; Jin, W.; Hu, W. Covalent surface modification of LY12 aluminum alloy surface by self-assembly dodecyl phosphate film towards corrosion protection. *Prog. Org. Coat.,* **2020**, *143*, 1-10, 105638.
[http://dx.doi.org/10.1016/j.porgcoat.2020.105638]

[25]   Rao, Y.; Wang , Q.; Oka, D. On the PEO treatment of cold sprayed 7075 aluminum alloy and its effects on mechanical, corrosion and dry sliding wear performances thereof, Surface & Coatings Technology **2019**, 383-1-58.
[http://dx.doi.org/10.1016/ j.surfcoat.2019.125271] [PMID: 125271]

[26]   Zhang, K.; Yu, S. Preparation of wear and corrosion resistant micro-arc oxidation coating on 7N01 aluminum alloy. *Surf. Coat. Tech.,* **2020**, *388*, 1-33, 125453.
[http://dx.doi.org/10.1016/j.surfcoat.2020.125453]

[27]   Huang, L.; He, L.; Chen, S.; Chen, K.; Li, J.; Li, S.; Liu, W. Effects of non-isothermal aging on microstructure, mechanical properties and corrosion resistance of 2A14 aluminum alloy. *J. Alloys Compd.,* **2020**, 1-45, 155542.
[http://dx.doi.org/10.1016/j.jallcom.2020.155542]

[28]   Shen, Y.; Dong, Y.; Yang, Y.; Li, Q.; Zhu, H.; Zhang, W.; Dong, L.; Yin, Y. Study of pitting corrosion inhibition effect on aluminum alloy in seawater by biomineralized film. *Bioelectrochemistry,* **2020**, *132*, 107408.
[http://dx.doi.org/10.1016/j.bioelechem.2019.107408] [PMID: 31816577]

[29]   Guan, F.; Duan, J.; Zhai, X.; Wang, N.; Zhang, J.; Lu, D.; Hou, B. Interaction between sulfate-reducing bacteria and aluminum alloys—Corrosion mechanisms of 5052 and Al-Zn-In-Cd aluminum alloys. *J. Mater. Sci. Technol.,* **2020**, *36*, 55-64.
[http://dx.doi.org/10.1016/j.jmst.2019.07.009]

[30]   Wang, J.; Xiong, F.; Liu, H.; Zhang, T.; Li, Y.; Li, C.; Xia, W.; Wang, H.; Liu, H. Study of the corrosion behavior of Aspergillus niger on 7075-T6 aluminum alloy in a high salinity environment. *Bioelectrochemistry,* **2019**, *129*, 10-17.
[http://dx.doi.org/10.1016/j.bioelechem.2019.04.020] [PMID: 31075534]

[31]   A., RagabKh.; Bouaicha, A. Development of fatigue analytical model of automotive dynamic parts made of semi-solid aluminum alloys *Science press,* **2018**, *28*, 1226-1232.

[32]   Wang, Z-Y.; Zhang, T.; Li, X.; Wang, Q-Y.; Huang, W.; Shen, M. Characterization of the effect of CFRP reinforcement on the fatigue strength of aluminium alloy plates with fastener holes. *Eng. Struct.,* **2018**, *177*, 739-752.
[http://dx.doi.org/10.1016/j.engstruct.2018.10.010]

[33]   Takahashi, K.; Osedo, H.; Suzuki, T.; Fukuda, S. Fatigue strength improvement of an aluminum alloy with a crack-like surface defect using shot peening and cavitation peening. *Eng. Fract. Mech.,* **2018**, *193*, 151-161.
[http://dx.doi.org/10.1016/j.engfracmech.2018.02.013]

[34]   Araújo, L.M.; Ferreira, G.V.; Neves, R.S.; Malcher, L. Fatigue analysis for the aluminum alloy 7050-t7451 performed by a two scale continuum damage mechanics model. *Theor. Appl. Fract. Mech.,* **2020**, *105*, 1-27, 102439.
[http://dx.doi.org/10.1016/j.tafmec.2019.102439]

[35]   Anis, S.F.; Koyama, M.; Hamada, S.; Noguchi, H. Mode I fatigue crack growth induced by strain-aging in precipitation-hardened aluminum alloys. *Theor. Appl. Fract. Mech.,* **2019**, *104*, 1-44, 102340.
[http://dx.doi.org/10.1016/j.tafmec.2019.102340]

[36]   Baragetti, S.; Borzini, E.; Božic, Ž.; Arcieri, E.V. On the fatigue strength of uncoated and DLC coated 7075-T6 aluminum alloy. *Eng. Fail. Anal.,* **2019**, *102*, 219-225.
[http://dx.doi.org/10.1016/j.engfailanal.2019.04.035]

[37]   Ding, Y.; Lv, Y.; Zhao, B.; Han, Y.; Wang, L.; Lu, W. Response relationship between loading condition and corrosion fatigue behavior of nickel-aluminum bronze alloy and its crack tip damage mechanism. *Mater. Charact.,* **2018**, *144*, 356-367.
[http://dx.doi.org/10.1016/j.matchar.2018.07.033]

[38]   Zou, Y.; Cao, L.; Wu, X.; Wang, Y.; Sun, X.; Song, H.; Couper, M.J. Effect of ageing temperature on microstructure, mechanical property and corrosion behavior of aluminum alloy 7085. *J. Alloys Compd.,* **2020**, *823*, 1-10, 153792.
[http://dx.doi.org/10.1016/j.jallcom.2020.153792]

[39]   Seisenbacher, B.; Winter, G.; Grün, F. Modelling the effect of ageing on the yield strength of an aluminium alloy under cyclic loading at different ageing temperatures and test temperatures. *Int. J. Fatigue,* **2020**, *137*, 1-26, 105635.
[http://dx.doi.org/10.1016/j.ijfatigue.2020.105635]

[40]   Khan, M.A.; Wang, Y.; Anjum, M.J.; Yasin, G.; Malik, A.; Nazeer, F.; Zhang, H. Effect of heat treatment on the precipitate behaviour, corrosion resistance and high temperature tensile properties of 7055 aluminum alloy synthesis by novel spray deposited followed by hot extrusion. *Vacuum,* **2020**, *174*, 1-5, 109185.
[http://dx.doi.org/10.1016/j.vacuum.2020.109185]

[41]   Das, U.; Toppo, V. Effect of Tool Rotational Speed on Temperature and Impact Strength of Friction Stir Welded Joint of Two Dissimilar Aluminum. *Mater. Today,* **2018**, *5*, 6170-6175.

[42]   Lu, J.; Song, Y.; Hua, L.; Zhou, P.; Xie, G. Effect of temperature on friction and galling behavior of 7075 aluminum alloy sheet based on ball-on-plate sliding test. *Tribol. Int.,* **2019**, *140*, 1-12, 105872.
[http://dx.doi.org/10.1016/j.triboint.2019.01.037]

[43]   Wang, J.T.; Zhang, Y.K.; Chen, J.F.; Zhou, J.Y.; Luo, K.Y.; Tan, W.S.; Lu, Y.L. Effect of laser shock peening on the high-temperature fatigue performance of 7075 aluminum alloy. *Mater. Sci. Eng. A,* **2017**, *704*, 459-468.
[http://dx.doi.org/10.1016/j.msea.2017.08.050]

CHAPTER 8

Corrosion of Nuclear Waste Systems

K. Ramakrishna Reddy[1], N. Suresh Kumar[2,*], K. Chandra Babu Naidu[3], B. Venkata Shiva Reddy[3,4] and T. Anil Babu[3]

[1] *Department of Chemistry, Reva University, Bangalore 560064, Karnataka, India*

[2] *Department of Physics, JNTU College of Engineering Anantapur, Anantapuramu-515002, A.P., India*

[3] *Department of Physics, GITAM Deemed to be University, Bangalore - 562163, Karnataka, India*

[4] *Department of Physics, The National College, Bagepalli-561207, Karnataka, India*

Abstract: In this chapter, the corrosion problems of nuclear waste systems in view of the disposal are discussed. The main form of waste of packages is discussed, ascertaining high-level waste, and the cemented intervening level radioactive waste forms, vitrifying nuclear waste, canister waste forms and nuclear waste glasses. The discussion between the rate of corrosion of all the nuclear waste packages with the nuclear waste disposal concept and the safety measures of the landfill sites is featured. Furthermore, the corrosion of the various kinds of nuclear waste packages and the metallic container for the high-level waste packages are reviewed. In view of the deterioration or dissipation processes, the experimental in-situ approaches, and the exemplary corrosion of nuclear waste and lifetime forecasting are discussed. The major challenge in global research is acquiring data and authentic forecasting for the function over long term periods, as landfill of these kinds of nuclear waste packages must abide safe for very long periods of years.

Keywords: Corrosion Issues, High-Level Waste, Nuclear Waste Forms, Radioactive Waste.

1. INTRODUCTION

Radioactive waste is defined as the material that is contaminated with radioactive nuclide at concentrations greater than a safe level. It does not have any practical purpose. Radioactive wastes are usually by-products of nuclear power generation and other applications of nuclear fission or nuclear technology, such as research and medicine. Radioactive waste is hazardous to most forms of life and the environment. It is regulated by government agencies to protect human health and

---

* Corresponding author N. Suresh Kumar: Department of Physics, JNTU College of Engineering, Anantapur-515002, Andhra Pradesh, India; Tel: +91-81211 27157; E-mail: sureshmsc6@gmail.com

N. Suresh Kumar, P. Banerjee, H. Manjunatha and K. Chandra Babu Naidu (Eds.)

the environment. That is, all waste materials which are radioactive in nature are called radioactive waste. The petrified issue is caused due to the changes in earth's temperature or changes in climate. It is authorized for the use of nuclear energy, such as the non-production of greenhouse gases. In further coming years, nuclear power extensively relies on the management of nuclear waste, produced at various steps of the nuclear fuel cycles. It is believed that the radioactive waste disposal across the world of high-level nuclear waste paralyzes them to a matrix of the solid (glass). The glass can disintegrate many elements existing in the periodic table, and it can be utilized as a medium for waste confinement [1, 2]. Glass is an exactly non-religious compound. All the elements present in the same compound are the main components in the glass consisting of the nuclear waste. With the high percolate resistance of the glass, it is impossible to disintegrate into water. Low percolate character is the required criteria for a solid compound to vitrify nuclear waste.

The poor corrosion inclination will identify the resultant nuclear waste glass. Borosilicate and lead iron phosphate glasses are appraised as a selective solid matrix for high-level nuclear waste [3, 4]. The ultimate disposal of high-level nuclear waste presumes multi-barrier approach. The radionuclide, which is coming from the nuclear active waste glass enters the biosphere. The groundwater encounters the nuclear active waste glass after perforating all the barriers applied for the nuclear active waste glass. The behavior of corrosion of nuclear waste glass always explains opposing percolate in the presence of moisture [5]. Knowing the percolation of nuclear waste, glasses are essential since they treated as a geological warehouse for a large amount of time. To find out directly the corrosion of long run is quite tough using small experiments. But for the performed experiments in laboratory, the long run corrosion can be measured.

The trustworthy disposal of long-living radioactive waste for geological time scale is one of the most important issues in the safety application and acceptance of nuclear energy. The trustworthy disposal involves the development of robust disposal concepts, explaining their safety over the geological chronicles, and knowing the properties of different natural barriers in the topographical warehouse. It took almost 30 years to do research to find performance all over the world into different nuclear waste forms, to study their properties like thermal, radiation, chemical, mechanical, and physical properties related to eventual long-term disposal. The main pattern used to paralyze nuclear waste are cement, bitumen, glass, and, more recently, ceramics and spent fuel. This chapter initiates the corrosion property of some of the major waste forms considered globally for a long period and high-level radioactive waste. High-level radioactive waste is glass, and intermediate-level waste is cement. Both are discussed since these are the most characteristic waste forms examined today. The corrosion of spent fuel is also reviewed since direct disposal of spent fuel is presently being examined in many places globally. The corrosion of the metallic container material is explained as well, as their part is very appreciative to the role of the waste form itself. Both constitute the waste package corrosion property, and it is one of the most predominant properties of the nuclear waste package. It contains the release of the radionuclides into the surrounding repository. Further, it governs to a large extent of lifetime of these engineered barriers in the repository.

The nuclear corrosion of glass of borosilicate in aqueous can be studied. The dissolution of the glass is the huddle of various processes and mechanisms. Dissolution began with the hydration of glass by the diffusion of water into the glass matrix. Further, the ion exchange process among the protons available in the earth crust and base metal in the production of glass occurs. The dissolution of glasses can be done by reacting with water since the ionic covalent bond is the most soluble element of the glass matrix.

2. POTENTIAL CORROSION ISSUES IN NUCLEAR WASTE PACKAGES

2.1. Description of Typical Waste Packages

The radioactive waste is generated in various extents of the nuclear fuel cycle. In the operation of mining of uranium ore, fuel generation, the operation of the nuclear power plant, handling of the spent nuclear fuel, and dismantling and decommissioning of the nuclear power plant at the end of its lifetime. Nuclear waste is also produced in research laboratories, hospitals, universities, chemical industries, *etc*. All these various types of nuclear wastes are subsequently handled (conditioned, immobilized) in such a way that the resulting waste packages can be safely stored and eventually disposed in surface or geological disposal sites [6]. The most popular kinds of radioactive waste packages and in those packages, the corrosion issue is important. Those are the corrosions of vitrified high-level waste, spent uranium oxide fuel packages, and cemented waste packages.

2.2. Vitrified High-Level Waste

Vitrified waste form immobilizes the sludge issuing from the reprocessing of spent fuel, and contains fission products, actinides, and activation products. Through a verification process, these highly radioactive solutions are incorporated in a homogeneous borosilicate glass matrix. In the French AREVA R7T7 process, for example, the glass is poured into stainless steel canisters of 134 cm height and 43 cm diameter. In that process, the radioactivity inventory per waste package is typically between $10^{15}-10^{16}$ Bq (β/γ) and $10^{14}-5 \times 10^{14}$ Bq (α), for a total content of fission product oxides, metallic particles, and actinide oxides of $\sim$14 wt %. Because of this very high level of radioactivity, it will take a minimum of 50 years before the outer temperature of the canisters drops below 100°C. Therefore, they can be disposed in a geological repository. A waste package consisting of high-level glass in its canister is further inserted in a mechanically and corrosion-resistant container (also called 'overpack'), ready for eventual geological disposal. This is because of the primary canister is not considered a performant barrier during the thermal period. Full details of the chemical composition and radiochemical inventory of some reference HLW glasses can be found in Lutze [7], Godon *et al.* [8, 9] and Caurant *et al.* [10].

2.3. Cemented Radioactive Waste

Generally, cemented materials are used to immobilize a large variety of nuclear waste, ranging from sludge's small particles (ashes, resins, *etc.*), metallic pieces or even large compacts [6, 11 - 13]. Immobilization in cement, in fact, consists of encapsulation, unlike verification which includes the uptake of the waste constituents at an atomic scale into the glass matrix. These wastes may originate from various stages of the back end of the nuclear fuel cycle, reactor operation, reprocessing, dismantling operations, or operations in other facilities (research, hospitals, *etc.*). They can be either homogeneous or heterogeneous. The levels of radioactivity of these wastes may be diverted and usually correspond to low-level or intermediate-level wastes [11].

The important constituent of the cement matrix is Ordinary Portland Cement (OPC), and blending agents such as blast furnace slags, and coal combustion fly ash are often added to enhance density and strength. The main components of cement consist of a gel like Ca silicate and portlandite. Sand–cement mixes, or cement containing sand and coarse aggregates, are also used in the cementation of radioactive waste [14]. It provides an update of the knowledge on the various properties of cemented waste. The cemented nuclear waste produced in this path is surrounded by a container, being itself either made of cement, concrete or metal. The research is being performed increasingly on the cement used as backfill or for engineered barriers in various kinds of repositories, covering issues such as gas generation and permeability, and the impact of saturation or non-saturation of the near field.

2.4. Corrosion Issues in Canister Material of Carbon Steel and Cast Iron

Generally, the composition of Carbon steel is alloy Fe and C having the content of Carbon less than 2 wt%, usually containing Manganous less than 1.65 wt%,

Silicon less than 0.60 wt % and copper less than 0.60 wt % also very small amount of various alloying metals such as Chromium, Nickel, Molybdenum, Tungsten, Vanadium, and Zirconium [15]. Very less content of C and mild steels are the types of alloys suggested being used for canister materials having carbon contents of 0.15 wt % and 0.29 wt %, respectively. Generally, cast iron is suggested as a structural material for a Cu outer shell containing an alloy of Fe and carbon greater than 2 wt % carbon and 1.3 wt % silicon. Usually, both materials like Carbon and mild steel are suggested for canister materials across the globe.

The carbon-steel has a wide variety of uses like canister material along with good properties of hardness and malleability, adequate corrosion effects in the anticipated archive environments, substantial experience with manufacturing and sealing large cylindrical repositories with profusion, very cheap. The drawback of carbon steel is the evolution of hydrogen and ferrous under anaerobic conditions which may severely affect the other boundaries, specifically Hillman Composite Beam. C-steel is categorized as a corrosion-acceptance compound and will corrode throughout material at a specific rate. It will be based on the redox property, pH, temperature, and anatomy of the aqueous phase. C-steel is not satisfactory to be used in persistently aerobic circumstances because of the viability of confined corrosion and the evolution of hostile ferric species that can affect the nuclear waste form or other barriers of the materials. In the presence of anaerobic circumstances, the rate of corrosion slowly is reduced with time since the formation of a passive layer of the product [16], in due course achieving a steady-state rate. This is changing with pH and the environment of the reaction condition in bulk solution or in compacted Bentonite is clearly explained in reference [17]. The constant corrosion rate is expected in Bentonite-backfilled archives. The rate of corrosion of cast iron is the same as that of carbon steel clearly elucidated in reference [17]. In the compound of cement or in caustic solutions replicating cement pore water of the steady-state corrosion rate of carbon steel is less than 0.1 mm/year [18].

The confined corrosion rate of carbon steel may take place because of confined disintegration of a corrosion-resistant layer leading to corrosion, irregular wetting of the surface since compression of surface contaminants, evolve in the geographical separation of cathodic and anodic areas and decreases dissolution of ferric sites corrosion products in the process of transition from aerobic to anaerobic conditions. The evidence from the reference [19] provides that carbon steel is not corrosion resistant in the relatively caustic pore water having the pH ≤ 9.5. This is an important exam due to the confined corrosion of carbon steel canisters, wherein the damage of a corrosion resistance layer should not produce. C-steel will be corrosion resistant in cemented surroundings and abide till the cement pore water persist caustic [20]. If the cement degrades and the pore-water pH declines, the surface will become permitting to confined corrosion at some juncture based on the strength of chloride. Presence of non-caustic surroundings and its confined attack produces the form of irregularity on the surface rather than varying pitting [19]. Caustic embrittlement is the form of stress corrosion cracking or different kinds of H_2-related cracking, including H_2-induced cracking and vesicle development [21, 22]. The different configurations of stress corrosion cracking are implausible to affect carbon steel canisters in archives since either the absence of the required surroundings chemically in the instances of phosphate, nitrate, and caustic cracking or the absence of cyclic loading in the instances of the suitable pH and large pH configurations of cracking that are perceived for the channel of steels in carbonate or bicarbonate surroundings [23].

2.5. Substances for Radioactive Waste Disposals

Since then, there are two important perspectives that are identified to strongly satisfy the essential force on the nuclear waste from over pack the corrosion-resistance and the corrosion acceptance hypothesis. Generally, the corrosion acceptance conviction compounds like carbon steel, copper in oxidizing chemical environments corrosion takes place at a consequential but low and moderately expected rate. Those materials would be used when enough thickness is allowed for the severe corrosion expected during the time. In the passive nature and moderately thin and extensive passive materials are based on the suggestion of the suitable container. Passive materials like stainless steel, nickel-based alloys and Ti alloys, Cu in reducing environments free of complexing agents, provide passive nature offensive in the anticipated disposal environments. Those compounds deteriorate at a very low rate and therefore, they can be used in relatively small opacity. By the usage of passive compounds, the high rate of risk of confined corrosion must be taken into consideration since this can lead to severe and random increases in the rate of corrosion.

Assessing the suggestible compound for the container materials of the disposal of high-level waste in deep geological formations and forecasting the corrosion behavior of the waste containers of a very long time of periods are reported across the globe disputes from the compounds of very long back from corrosion points of view. In terms of the French concept, the overpack not only produces as a part of the coherence hurdles but is also a prime component of the reversibility that is suggested for the French geological archive. Reversibility means the probability of restoring positioned packages as well as arbitrating and changing the disposal process and design. Everlasting safety and reversibility are the main accompanying principles that are led to the basic layout of the geological archive in an argillaceous production. The above concept consolidates high safety from the beginning phases of the design and permits the continuous inclination of choices towards solutions providing the highest lustiness with respect to knowing their unpredictability and proposing blockage and conservation quantities against identified threats.

2.6. Nuclear Waste Forms of The Glasses

Across the globe, the glasses are identified as the most suitable material for the immobilization of both high-level waste and low-level waste by the underground repository. In addition to the chemical durability, mechanical integrity and thermal stability, glasses are flexible with waste loadings and possess the capability of incorporating most of the waste elements. The method of process of introducing high-level waste by chemically dissolving them into glasses is known as vitrification. Currently, glasses are used on an industrial scale for the immobilization of high-level wastes. Since the ability of glass to incorporate ions with a wide variety of radii and charges, large numbers of nuclear waste glasses are also investigated for immobilization. Considering the ensuing anatomy remains undisturbed, the method by which radionuclides can reach the biosphere. It will be dissolved by the waste form in aquifer water, accompanied by the relocation of the nuclear solution to the surface. They accompany that one of the important factors in sort outing a nuclear waste compound for ensuing the disposal in an underground archive for immune to leaching by under earth crust water that may lead to perforate the surroundings. Because of the above reason, a huge number of glasses are studied over a long period of time for the disposal of high-level waste.

These glasses are generally metastable in nature related to time and temperature [24]. Glass materials possess high thermal stability, which is one of the important prosperities. The glass should be adequate stable to empower the glass to be tempered without the formation of the crystallization process. Usually, the glasses will crystallize if we heat to adequate temperature over a certain time of period. Generally, this crystallization happens in an unconfined process by nucleation of the huge number of crystals from external surfaces. This causes the production of unpredictable stresses to lead to the cracking of the glass. The steady state temperature is especially very much essential in the case of glasses to be used for paralyzing high-level waste since the low stability tends to the production of crystals either in the process of annealing of the glass or while in the time of storing. If severe cracking of the glass happened continuously, the discharging rate would be remarkably increasing.

2.7. Borosilicate Glass Waste Form

Borosilicate glasses are generally suggestible waste forms for the immobilization of high-level waste since they voluntarily diffuse into various ranges of waste compositions. Moreover, they can be easily altered with respect to upgrading the characteristics of materials. Furthermore, they form the basis of the commercial glass industry and are studied substantially over the period. Therefore, those are characterized very significantly, and the properties of the materials are well known and appreciable. Generally, the high durable waste form will be vitreous silica; however, they need huge processing temperature. The materialistic glass configurations will be found through glass robustness, processing potentiality and productivity. Usually, boric acid is most preferred to influence the change in behavior of silica. This significantly reduces the operating temperature required to produce glass and its feasibility. However, conserving good robustness in the range of suggested composition. Since a very large number of studies happen, borosilicate glass becomes the selective material across the globe for the immobilization of radioactive waste materials. The choice of the material depending on the malleability of borosilicate glass tends to the filling of nuclear waste and the capability to integrate various kinds of nuclear waste metals, accompany with good glass fabricating capability, chemical resistance, mechanical strength, and significant temperature and radiation stability. Chemical constituents are studied usually concerted on the sodium borosilicate compound with few inclusions of different kinds of oxides, includes Al_2O_3, Li_2O, CaO, ZnO, *etc*. Now a huge amount of database is developed for the borosilicate glasses to paralyze the high-level waste with substantial details accessible on the transforming properties and corrosion character, mechanical strength, temperature stability, in addition to crystallization character, and radiation stability [25, 26].

2.8. Phosphate Glass Waste Form

An extensive report is available for freezing of high-level waste based on borosilicate composition [27 - 30]. The properties of chemical bonding of phosphate glasses in different aspects resemble the organic polymers rather than silicate networks. Furthermore, it tends to various characteristics between the two types of glasses. Phosphate glasses possess very fewer melting points, poor melting viscosities and considerable changes in temperature viscosity behavior. Huge amount of $NaH_{14}Al_3(PO_4)_8 \cdot 4H_2O$, iron alumina-phosphate and zinc-phosphate compositions are eventually produced and provide better chemical resilience. However, the stability of their temperatures is still relatively poor, and they are severely corrosive. Very recently, the group of lead iron phosphate glasses are being introduced at "Oak Ridge National Laboratory" in the United

States of America. The lead iron phosphate glasses manifested the significant glass-forming property, both having good temperature stabilities and very good chemical resistance. Furthermore, the latest melts are not at all corrosive like a phosphate glass. These glasses will be synthesized at a higher temperature range of 800-1000°C. This temperature range is lesser than the borosilicate glasses which are presently diligent for immobilization of HLW. However, it is proved already that a minimum operating temperature is 1000°C tends to dissolve HLW completely. However, discharge rates are reported to be 10-1000 times lesser than the borosilicate glasses if the crystallization of phosphate glasses is permitted occurring the reduced robustness. Furthermore, resilience in aqueous solutions also reduces remarkably at 100°C. This is important while in view of repository surroundings where the temperatures may appreciably cross the value. It proves the increasing composition of ferric oxide of the glass. As decreasing the content of lead oxide, the glass resilience is enhanced, especially in caustic solutions. However, more research work must be done in this domain while claiming definitive conclusions. Moreover, it is noted that calcium oxide provides more durability to iron phosphate glasses.

2.9. Waste Form of Rare Earth Oxide Glass

Very fewer results are published in nuclear waste form of rare earth metal oxide glasses for immobilization applications. However, materialistic lanthanide borosilicate compound is identified. These materials are described as Loffler glasses and they are being introduced as optical glasses since 1930 and possessing 55 wt % of oxides of lanthanides. Very recently, it is proposed as a significant host for the immobilization of uranium, plutonium and americium showing very good solubility than the regular borosilicate glasses.

2.10. High Silicate Glass Waste Form

Pure SiO_2 glasses possess better discharge resistances than any other borosilicate glasses. However, they are melted at higher temperatures. Alumino-silicate glasses reported long back of waste vitrification requiring the higher melting temperatures than the borosilicate compositions. In the year 1960, the Alumino-silicate glasses possessing the active waste is buried in the Chalk River in Canada and it was kept track till 1978 when it is removed completely. This process gives information that only reported material of immobilized active vitreous waste is being monitored. It is calculated by the data available in the process that it would be twenty million years for the complete dissolution under specific conditions. Alumino-silicate glasses are deliberated of late by Vance *et al.* [31]. It is suggested that up to 20 wt % uranium dioxides could be retained when the extensive cooling of melt occurs. However, 10 wt % was held on if the rate of cooling is 5°C min^{-1} from 1400°C. Huge melt temperatures (1600°C) are required for these glasses since it contains the high silica. The resilience is suggested to be very high when comparing with borosilicate glasses. It is reported that alkali alumina-silicate glass is also a potential second-generation waste form with improved durability [32].

CONCLUSION

The nuclear corrosion waste systems of different types like cemented intervening level radioactive waste forms, vitrified nuclear waste, canister waste forms, nuclear waste glasses, rare earth oxides are discussed. These are about working of nuclear waste system compositions and its disposal methods. Furthermore, the corrosion of the various kinds of nuclear waste packages, and the metallic container for the high-level waste packages are reviewed in view of the deterioration of dissipation processes, the exemplary of corrosion of nuclear waste and lifetime forecasting. The major challenge in this globally oriented research is acquiring data and authentic forecasting for the function over long term periods, as landfill of these kinds of nuclear waste packages must abide safe for long time periods of years.

CONSENT FOR PUBLICATION

Not applicable.

CONFLICT OF INTEREST

The author declares no conflict of interest, financial or otherwise.

ACKNOWLEDGEMENTS

Declared none.

REFERENCES

[1]   Mukerji, J.; Sanyal, A.S. Historical series: Indian work on vitreous matrices for the containment of radioactive waste 1960-1980. *Glass Technol.,* **2004**, *45*, 117-125.

[2]   Raman, S.V. The effect of mixed modifiers on nuclear waste glass processing, leaching and Raman spectra. *J. Mater. Res.,* **1998**, *13*(1).
[http://dx.doi.org/10.1557/JMR.1998.0002]

[3]   Pankou, A.S.; Batyukhova, O.G.; Ojoven, M. I and Lee W. E, Simulation of Self-Irradiation of High-sodium content Nuclear Waste Glasses. *Proc. MRS,* **2007**, 985.

[4]   Jain, V; Pan, Y.M High-Level waste glass dissolution in simulated internal waste package environments. *Scientific research on the back-end of the fuel cycle for the 21 century,* **2000**, 575.

[5]   Yanagi, T.; Yoshizoe, M.; Nakatsuka, N. *Leach Rates of Lead-Iron Phosphate Glass Waste Forms,*

Osaka University WM'02 Conference, **2002**, pp. 24-28.

[6]   Ojovan, M.I.; Lee, W.E. *An Introduction to Nuclear Waste Immobilisation*; Elsevier, **2005**.

[7]   Lutze, W.; Erwing, R. Silicate glasses. In: *Radioactive Waste Forms for the Future*; North Holland: Amsterdam, **1988**; p. 791.

[8]   Godon, N *Long term behavior of nuclear waste glass, CEA Technical Report,* **2002**.

[9]   Godon, N.; Gin, S.; Minet, Y.; Grambow, B.; Lemmens, K.; Aertsens, M. Reference report on the state of the art of glass properties and glass alteration during long term storage and under disposal conditions *Deliverable 1.1.1 of RTD component 1, Part I. NF-PRO project with the European commission,* **2005**.

[10]   Caurant, D.; Loiseau, P.; Majerus, O.; Aubin-Chevaldonnet, V.; Bardez, I. Glass-Ceramics and Ceramics for Immobilization of Highly Radioactive Nuclear Wastes. *Nova Sci.,* **2007**.

[11]   Improved cement solidification of low and intermediate level radioactive wastes *IAEA TRS-350,* **1993**.

[12]   Status of technology for volume reduction and treatment of low and intermediate level solid radioactive waste *IAEA TRS-360,* **1994**.

[13]   Glasser, F.P.; Atkins, M. Cements in radioactive waste disposal. *MRS Bull.,* **1994**, 33-38. [http://dx.doi.org/10.1557/S0883769400048673]

[14]   Wieland, E.; Johnson, C.A.; Lothenbach, B.; Winnefeld, F. Mechanisms and modelling of waste/cement interactions *Res. Soc. Symp.,* **2006**, pp. 663-672.

[15]   ASM Metals Hand book, Corrosion. *American Society for Metals International,* **1987**, *13*.

[16]   Smart, N.R.; Blackwood, D.J.; Werme, L.O The anaerobic corrosion of carbon steel and cast iron in artificial ground waters. *SKB Technical Report,* **2001**, 1-47.

[17]   Johnson, L.H.; King, F. The effect of the evolution of environmental conditions on the corrosion evolutionary path in a repository for spent fuel and high-level waste in Opalinus Clay. *J. Nucl. Mater.,* **2008**, 379. [http://dx.doi.org/10.1016/j.jnucmat.2008.06.003]

[18]   Smart, N.R.; Blackwood, D.J.; Marsh, G.P.; Naish, C.C.; O'Brian, T.M.; Rance, A.P.; Thomas, M.I. The anaerobic corrosion of carbon and stainless steels in simulated repository environments: A summary review of Nirex research. **2004**. AEAT/ERRA-0313.

[19]   H 12 project to Establish the Scientific and Technical Basis for HLW Disposal in Japan. *JNC,* **2000**.

[20]   Kursten, B.; Druyts, F.; Macdonald, D.D.; Smart, N.R.; Gens, R.; Wang, L.; Weetjens, E.; Govaerts, J. Review of corrosion studies of metallic barrier in geological disposal conditions with respect to Belgian super container concept. *Corros. Eng. Sci. Technol.,* **2001**, *46*, 91-97. [http://dx.doi.org/10.1179/1743278210Y.0000000022]

[21]   King, F. Hydrogen effects on carbon steel used fuel containers. *Nuclear Waste Management Organization Report,* **2009**, 29.

[22]   Turnbull, A. A review of the possible effects of hydrogen on lifetime of carbon steel nuclear waste containers, national cooperative for the disposal of radioactive waste. *Nagra Technical Report,* **2009**.

[23]   King, F. Container materials for the storage and disposal of nuclear waste. *Corrosion,* **2013**, *69*, 986-1011. [http://dx.doi.org/10.5006/0894]

[24]   Gribble, N.R.; Short, R.; Short, R.; Urner, E.; Riley, A.D. The impact of increased waste loading on vitrified HLW quality and durability. *Proc. MRS,* **2009**, *1193*, 283-290. [http://dx.doi.org/10.1557/PROC-1193-283]

[25]   Cunnane, J.C.; Bates, J.K.; Bradley, C.R. *High level waste borosilicate glass: a compendium of corrosion characteristics*; United States Department of the Environment, **1994**, p. 2.

[26] Chick, L. A; Piepe:, G. F; Mellinger, G. B; May, R. P; Gray, W. J; Buckwalter, L.Q The effects of composition on the properties in an 11-component nuclear- waste glass system, Pacific Northwest Laboratory *Battelle* **1981**.

[27] Ojovan, W.E.; Lee, M. *New developments in glassy nuclear waste forms*; Nova Science Publishers, **2007**.

[28] Vashman, A.A.; Samsonov, V.E.; Demin, A.V. *Phosphate Glasses with Radioactive Waste*; CNIIatominform, Moscow, Russia, **1997**.

[29] Vance, E.R.; Urquhart, S.; Anderson, D.; George, I.M. Advances in ceramics, Nuclear waste management II. *American Ceramic Society Westerville,* **1986**, *20*, 249-258.

[30] Gahlert, S.; Ondracek, G. *Radioactive waste forms for the future*; North-Holland: Amsterdam, **1988**, pp. 161-192.

[31] ROSS W. A. Sintering of radioactive wastes in to a glass matrix, Pacific Northwest Laboratory *Battelle,* **1975**, 32.

[32] Gahlert, S.; Ondracek, G. *'Sintered galss' in Radioactive waste forms for the future*; Lutze, W.; Ewing North-Holland, R.C., Eds.; Amsterdam, **1988**, pp. 161-192.

CHAPTER 9

Microbiologically Influenced Corrosion

K. Ram Mohan Rao[1,*], K. Haripriya[2], P. Banerjee[3] and **A. Franco[4]**

[1] *Department of Chemistry, GITAM Deemed to be University, Visakhapatnam 530045, Andhra Pradesh, India*

[2] *Department of Anaesthesia (prev.), MIMS Medical College, NTR University, Vijaywada, Andhra Pradesh, India*

[3] *Department of Physics, GITAM Deemed to be University, Bangalore Campus, Visakhapatnam 530045, Andhra Pradesh, India*

[4] *Instituto de Física, Universidade Federal de Goiás, Goiânia, Brazil*

Abstract: Microbiologically influenced corrosion (MIC) is the subject of concern in various fields like industries related to healthcare, marine, petroleum, oil, *etc*. An attempt is made to present MIC and underlying mechanisms. Cathodic depolarization theory along with the other mechanisms supporting MIC caused by sulfate/nitrate-reducing bacteria is the focus of this chapter. Another important aspect of preventive and mitigation measures of MIC is concerned.

Keywords: Anaerobic, Biofilm, Mechanisms, Microbiologically Influenced Corrosion.

1. INTRODUCTION

Microbiologically Induced Corrosion (MIC) is one of the major problems affecting the economy and ecological safety of various sectors that rely on steel structures. MIC is known to be caused by different types of bio-organisms, *e.g.*, bacteria, algae, and fungi. It causes a significant loss of materials and adds to the cost of damage due to corrosion. Microbes colonize on the surface of materials resulting in the formation of biofilms that are responsible for the infections and microbiologically influenced corrosion (MIC). Generally, multiple organisms are responsible for the formation of a biofilm or outgrowth on the structure causing localized corrosion. The cause and progress of MIC are not clearly understood, which vary in different environments like oil transmission pipelines, marine environment, *etc.* Axelsen and Rogne (1998) first identified MIC more than a

---

* **Corresponding author K. Ram Mohan Rao:** Department of Chemistry, GITAM Deemed to be University, Visakhapatnam 530045, Andhra Pradesh, India; Tel: +91- 91138 54549; Email: rammohanrao.k@gmail.com

N. Suresh Kumar, P. Banerjee, H. Manjunatha and K. Chandra Babu Naidu (Eds.)

hundred years ago [1]. It was first defined by Videla in 1996, and according to him, "MIC is an electrochemical process in which the microorganisms are present to initiate, facilitate and accelerate the corrosion reactions" [2]. Industrial areas like the healthcare industry, marine, oil and gas industry are affected seriously by MIC. Numerous other areas like water utility, *etc.*, are affected severely. According to Flemming (1996) and Videla (2002), MIC is responsible for the leakage of pipelines and the plugging of injection wells causing loss of production and safety issues [3, 4]. Watkinson *et al.* (1984) explained that the condensed water collected at the bottom of the fuel tanks or pipes is responsible for microbial growth forming sediments, sludge, and slime. These are the elements that cause fuel deterioration and corrosion under the biomass produced [5].

Gaylarde *et al.* (1999) and Dzięgielewski *et al.* (2009) showed that fuel used in aviation and diesel fuel is prone to microbial growth because of the use of carbon microorganisms from C10-C18 hydrocarbon chains. At the interface of fuel and water, the nutrients and water are sufficient for the microbial activity and proliferation to take place, which results in the formation of biofilm. The colonization also takes place at the bottom and the surfaces of tanks and pipelines. At the bottom level of sludge and sediments, the obligate anaerobes, for example: sulfate-reducing bacteria (SRB), grow mostly in the anoxic zones, whereas the fuel phase contains only the aerobic microbes [6, 7]. In fuel turbidity, bottom sludge and sediments, unpleasant odour is caused by microbial contamination. The contaminants *via* distribution and transport systems pass to several locations. Further, they form the metabolic byproducts. Filter plugging, pinhole leaks of the pipeline systems and tanks are due to MIC, which is an electrochemical bio-corrosion led by the metabolites of the sulfate-reducing bacteria [8]. Biodiesel is a renewable source and used as an alternative source of fuel, which is non-toxic and contains the Fatty Acid Methyl Esters (FAME) as the main constituent and is more biologically active. The final blend of biodiesel is found to be contaminated by microorganisms [9 - 11]. It is found that not only sulfate-reducing bacteria (SRB) are responsible for MIC, but several other microorganisms are also responsible, *e.g.*, acid producers, general aerobic bacteria, iron oxidizers, iron reducers, manganese-oxidizing bacteria, methanogens, *etc.* [12]. Biomedical implants and corrosion are the subject of research for prolonging the life of implants and the cost of their replacement. For example, dental, cardiovascular, orthopaedic implants, *etc.*, are made up of stainless steel. The body fluid leads to corrosion of these stainless-steel implants, and the ions released due to corrosion are toxic [13]. Microorganisms are another reason for corrosion which live on the surface of body implants and cause infections. For example, implants fail most often by pitting and crevice corrosion [13 - 17]. For the body implants, AISI 304 and 316L stainless steels are the commonly used alloys that are known to be adversely affected by *P. aeruginosa*. It is found that in biologically active corrosion or microbiologically influenced corrosion, the major role is played by the accumulated microorganisms and the resulting biofilm on the surface of the materials. The role of biofilms in the mechanism of MIC is important to know how corrosion occurs on the surface of the materials and also how to combat or minimize corrosion.

2. A BRIEF HISTORICAL PERSPECTIVE

The first observation that microbially influenced corrosion occurs in metals was by Garett in 1891, who found that the lead sheathed cables undergo corrosion by bacterial metabolites [18]. In the year 1910, Gaines observed MIC and reported the involvement of bacteriogenic sulfur in corroding the internal and external water pipes [19]. Later, Ellis and Harder, in 1919, proved the presence of iron and sulfur bacteria in the deposits [20]. Until 1934, it was remained to unfold that why the underground structures undergo corrosion even though there is no stray current. Then in 1934, Kühr and Vlugt [21] studied graphitization of cast iron in anaerobic soils and found the anaerobic iron corrosion and sulfate reduction to form iron sulfide. Based on the observations, they proposed the first theory based on the bacterial depolarization of local cathodes with the involvement of anaerobic sulfate reducers. Some other groups accumulated the findings to emphasize the bacterial involvement of anaerobic sulfate reducers in inducing MIC [21 - 28]. In 1949, Skybalski and Olsen proposed mechanism that the aerobic MIC is followed by the formation of oxygen concentration cells due to the formation of tubercle [28]. Later, this mechanism was extended by Uhlig in 1953 [29], and the characteristics of steel were studied in the 1960s by other groups [29 - 38]. Iverson *et al.* [38] concentrated on the corrosion of steels by sulfate-reducing bacteria (*e.g.*, Desulfovibrio species).

Since 1980, increasing attention is drawn by the industrial sector and many MIC related problems are studied by several workers throughout the world. The evidence is the existing literature on the complex interaction of the metal-biofilm interface and microorganisms and the material characterization [39 - 43]. In the 21st century, the work was devoted to the characterization and analysis of complex microbiological organisms. It was found that in the marine and petroleum reservoirs, the SRB is not the sole culprit, but other microorganisms, *e.g.* those found in industries, also influence MIC [44 - 46]. Some other microorganisms later found were nitrate-reducing bacteria (NRB), iron-oxidizing bacteria (IOB), sulfur-oxidizing bacteria, archaea, methanogens, *etc.* [47, 48]. It is shown by some groups that some of the microorganisms help to reduce corrosion. In the year 2009, Videla and Herrera observed the corrosion inhibition to be induced by microbiological organisms like iron-reducing bacteria [49, 50].

3. BIOFILM

Microbiologically influenced corrosion initiates under the biofilm which assembles the corrosive microbes. It was first observed by van Leeuwenhoek on the surface of the tooth [51]. Microorganisms accumulate on material surfaces to form biofilms. Wettability, electrostatic charge, *etc.*, of the material's surface is affected after the biofilm formation, which causes the colonization of microorganisms. Initially, microorganisms produce extracellular polymeric substances (EPS), which consist of water along with polysaccharides, protein, nucleic acids, and lipids [52, 53]. EPS supports the cells on the surface and acts as the protective barrier so that microbes resist antimicrobials and support the horizontal gene transfer to make the biofilm to adapt to harsh environments [54].

Corrosion is a function of solid/liquid interfacial phenomena; hence the kinetics depends on the Physico-chemical environment at the interface, *i.e.* concentration of the oxygen, pH, salts, redox potential, *etc.* These are affected by microorganisms growing at the interface. Metabolic activities of the biofilms can alter all these parameters and thus influence corrosion [55, 56]. The microorganisms consist of corrosive sessile cells and also the non-corrosive planktonic counterparts. Due to this, the assessment of total corrosive activity becomes difficult. In any case, once the biofilm is formed, it resists the diffusion of chemical species, and thus the differential oxygen concentration cells are developed. When the biofilm consists of voids or channels in the vicinity of micro-colonies, the oxygen diffusion becomes faster or may not be consumed by the microorganisms. The oxygen-depleted site, *i.e.* anaerobic under respiring colonies, becomes the anode and the surrounding acts as the cathode. If the biofilm does not cover the film completely in that case, also differential oxygen cells develop and corrosion initiates. In some of the cases the biofilms consist of the non-corrosive microbes, it can reduce corrosion by acting as the barrier layer, *e.g.*, Pseudomonas S9 and Serratiamarcescens EF 190 reduced corrosion of carbon steel [57 - 59]. The attachments of microbes depend on surface roughness and surface energy of the materials [60]. So, the favorite sites for the attachments of the microbes were found to be porous weld joints, grain boundaries, scratches, *etc.* These are also known as the sites to activate abiotic corrosion.

4. MECHANISMS OF MIC

Microbiologically influenced corrosion is an interdisciplinary subject area where the knowledge of chemistry, metallurgy, microbiology is required to understand and mitigate the problem. The mechanisms of MIC had already been proposed by several groups since the 1900s. In 1934, von WolzogenKuhr and van der Vlugt proposed the first classical theory known as the 'Cathodic depolarization theory

[21]. Later some other theories were put forward, but still, there is no clear mechanistic understanding of MIC. MIC can be divided into two types: (a) Aerobic and (b) Anaerobic.

a. **Aerobic:** MIC occurs when the concentration of dissolved oxygen is high enough to favor corrosion without reducing the water [61]. The mechanism of aerobic MIC is already presented in the literature [62].
b. **Anaerobic:** MIC is supported by the necessary water reduction condition required for the cathodic current for the corrosion reaction. It is thought to be more damaging and hence the economic loss.

These definitions depend on the metal, whether it is active or passive, *i.e.* rate of corrosion. For the same environment, it may be anaerobic if the metal is active, and in reverse, it is aerobic when the metal is passive. Apart from dissolved oxygen, the other reducible species also exist in the environment. So, it is not the universal definition. These definitions, however, are valid when the comparison is made based on corrosion potential, *i.e.*, if positive with respect to hydrogen electrode, it is said to be aerobic and *vice versa*. If the corrosion potential is positive, but the working condition is anaerobic, which is already reported, it could be due to the presence of residual oxygen favoring reduction reaction required for the cathodic current. In the anaerobic conditions, various mechanisms for MIC were proposed mostly based on the sulfate-reducing bacteria. However, it was not very satisfactorily explained. The important mechanisms are given in the following sections.

4.1. Cathodic Depolarization Theory

Cathodic depolarization theory on the basis of electrochemistry was first proposed to understand MIC [21, 63]. The source of the mechanism was SRB. It was proposed that the corrosion occurred because of the reduction of cathodic hydrogen by sulfate-reducing bacteria and its hydrogenase enzyme. From the principle of corrosion of metals, the removal of hydrogen favors the cathodic reaction and hence depolarizing the OCP (free potential/open circuit potential). This would result in increased corrosion kinetics [64, 65]. It is known that the corrosion in anoxic a neutral medium, where only the hydrogen reduction, *i.e.*, the electron-accepting or cathodic reaction, is the rate-determining step, the rate of corrosion is very slow. However, the rate of corrosion significantly increases when the environment is supporting the proliferation of sulfate-reducing bacteria. In this case, the rate of corrosion may increase 70 to 90 times more than sterile controlled experimental condition [66 - 68]. In the classical cathodic depolarization theory proposed by Kühr and Vlugt [21], corrosion of metals occur by following the reactions given as under:

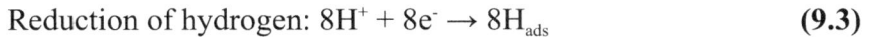

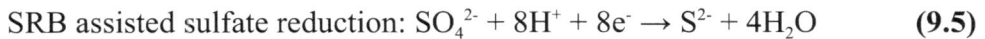

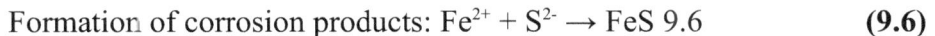

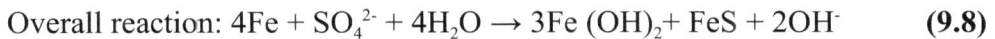

$$\text{Oxidation of iron: } 4Fe \rightarrow 4Fe^{2+} + 8e^- \tag{9.1}$$

$$\text{Water dissolution: } 8H_2O \rightarrow 8H^+ + 8OH^- \tag{9.2}$$

$$\text{Reduction of hydrogen: } 8H^+ + 8e^- \rightarrow 8H_{ads} \tag{9.3}$$

$$\text{Cathodic depolarization: } 8H_{ads} \ (4H_2) \rightarrow 8H^+ + 8 \tag{9.4}$$

$$\text{SRB assisted sulfate reduction: } SO_4^{2-} + 8H^+ + 8e^- \rightarrow S^{2-} + 4H_2O \tag{9.5}$$

$$\text{Formation of corrosion products: } Fe^{2+} + S^{2-} \rightarrow FeS \ 9.6 \tag{9.6}$$

$$3Fe^{2+} + 6OH^- \rightarrow 3Fe \ (OH)_2 \ 9.7 \tag{9.7}$$

$$\text{Overall reaction: } 4Fe + SO_4^{2-} + 4H_2O \rightarrow 3Fe \ (OH)_2 + FeS + 2OH^- \tag{9.8}$$

In the above reactions, HS^- dominates whereas S^{2-} only when the pH of the solution is very high; say around 10.

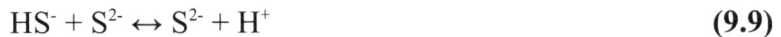

$$HS^- + S^{2-} \leftrightarrow S^{2-} + H^+ \tag{9.9}$$

In an acidic environment, H^+ ions are consumed and produce H_2S thereby increasing the pH of the environment [69]

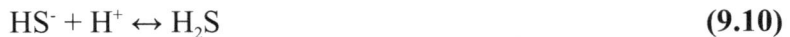

$$HS^- + H^+ \leftrightarrow H_2S \tag{9.10}$$

The adsorbed hydrogen (H) atoms on the surface of the cathode impede the cathodic reactions until freed from the surface. This process is known as cathodic polarization. As the desorption of H_{ads} requires high activation energy, the reaction steps are rate-limiting. A group of enzymes released from the SRB cells is called hydrogenase, which helps to catalyze the H_{ads} by lowering the activation energy and converts it to H_2 molecules and finally H^+ ions. This process is called the cathodic depolarization. Cathodic polarization theory is not proved to be a very successful theory in explaining the MIC as it suffers from several drawbacks. Biofilm, which has significant catalytic activity, has not been taken into consideration in this theory. The theory does not explain why corrosion is caused by hydrogenase- negative strains of SRB [70]. The effects of H_2S and the metabolites which are corrosive in nature are not considered in this theory [71]. Gu and Xu [72] suggested that the reduction occurs inside SRB cytoplasm not necessarily on any physical cathode.

Metals corrode because of the consumption of cathodic hydrogen, *i.e.*, the microbial uptake of hydrogen is the necessary condition for the corrosion. This means that the following equation must satisfy as:

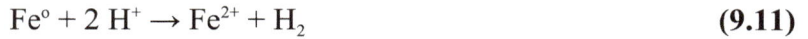

$$Fe^{\circ} + 2\,H^{+} \rightarrow Fe^{2+} + H_2 \tag{9.11}$$

and

$\Delta G^{\circ} = -10.6\ \text{kJ/mol}$

However, this classical theory is not satisfactory in all cases. Since the equation (9.11) is thermodynamically favored, in the environmental conditions, the anaerobic corrosion of Fe occurs even in the absence of microbes which are responsible for the H_2 uptake [73, 74]. It is the limited availability of hydrogen and slows kinetics of the H_2 formation slowing down the rate of anaerobic corrosion of iron. SRB consumes cathodic hydrogen as a substrate, however not affecting the corrosion of iron significantly [75, 76].

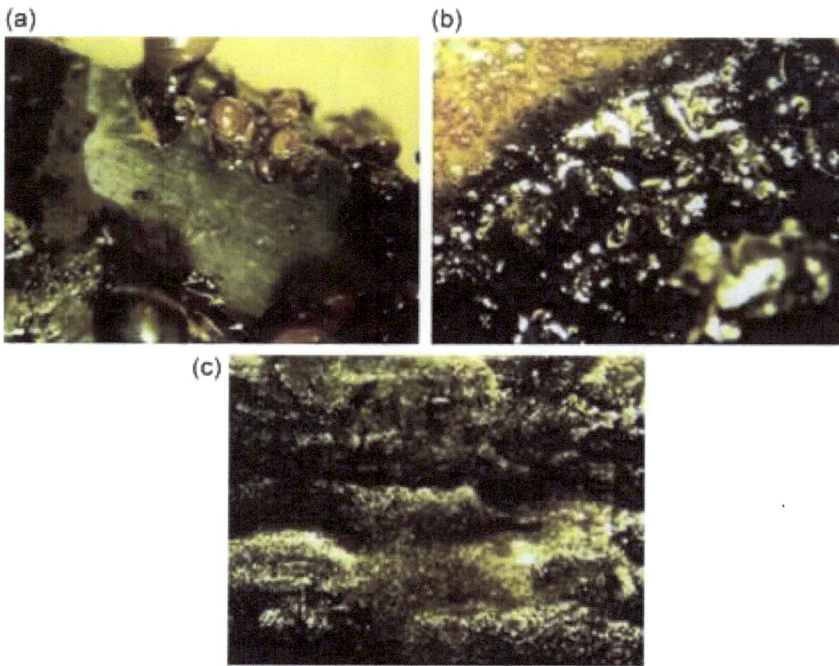

Fig. (1). Images of carbon steel after 3 months of exposure to the mixed population of microorganisms. **(a)** Tubercells after the activity of aerobic iron oxidizer microbes; **(b)** anaerobic Desulfovibrios caused the formation of shiny blackish biofilm and **(c)** under the blackish layer significantly roughened surface (mag. x 10) [51].

Emerging theories in sulfate-reducing bacteria-induced corrosion (EMIC) eliminates the above mentioned slow abiotic corrosion due to the hydrogen formation and involves the SRB to use iron more efficiently by direct uptake of electron released from the oxidation of iron [77]. Sulfate reducers by consuming the electron act as the driving force for the anodic dissolution of iron, as shown in Fig. (**1**). The kinetics of anodic dissolution at room temperature and in the absence of microbes which are seemed to be impossible is now favored. It is evident from Fig. (**2**) that the microbial oxidation of four moles of Fe to Fe^{2+} is accompanied by the reduction of only one mole of sulfate. The formed H_2S produces one mole of FeS and the left out three moles of Fe^{2+} produce iron minerals like iron carbonate which is abundant in produced water. Enning *et al.*, in 2012, suggested that the crust formed after microbial corrosion is electrically conductive. Hence, the contact of SRB with the metal surface if not necessary for driving the corrosion reaction [78]. In fact, the electrons released from the oxidizing Fe move to the cells attached to the crust deposited on the metal surface and reduced the sulfate according to the equation below:

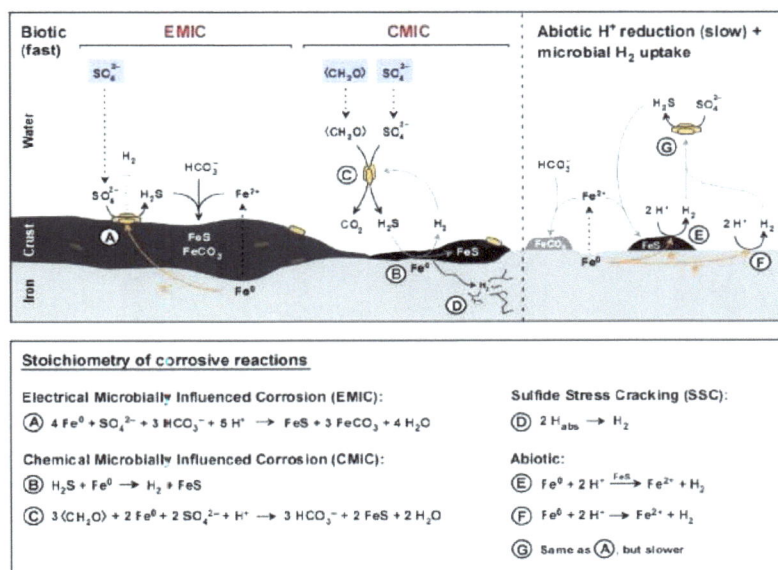

$$8e^- + SO_4^{2-} + 10\ H^+ \rightarrow H_2S + 4H_2O \qquad (9.12)$$

Fig. (2). Schematic representation of corrosion caused by sulfate-reducing bacteria. SRB corrodes iron *via* electrical microbially influenced corrosion or chemical microbially influenced corrosion [65].

Venzlaff *et al*. [74] also suggested that the requirement of SRB for the occurrence of corrosion. It is suggested by Enning *et al*. [78] that electrons flow to SRB through H_2S, whereas Dinh *et al*. [75], and Mori *et al*. [76], suggested from the similar electron uptake mechanism in some methanogenic archaea, where there is no H_2S formation in methanogenic culture and the corrosion of iron does not need the presence of H_2S.

Emerging microbial influenced corrosion, though involved only the highly corrosive SRB however, it is found that all SRB can influence corrosion by producing chemical H_2S. This is known as chemically and microbially influenced corrosion (CMIC) that occurs in the presence of sulfate and electron donors, *i.e.*, indirect catalyst of the anaerobic corrosion. From Fig. (3), it is evident that the CMIC occurs by degrading the organic matter by producing sulfide in anoxic environments. Apart from the corrosion by following microbial sulfate reduction, the CMIC also occurs by dissimilatory reduction of thiosulfate for sulfite, which produces corrosive sulfide significantly [79]. In Fig. (4), reaction B supports the intracellular oxidation of organic compounds by SRB produces sulfide and corrodes the iron. From Fig. (4), equation C can be used to calculate the amount of Fe corroding by consumption of a certain amount of acetate in the presence SRB [80, 81].

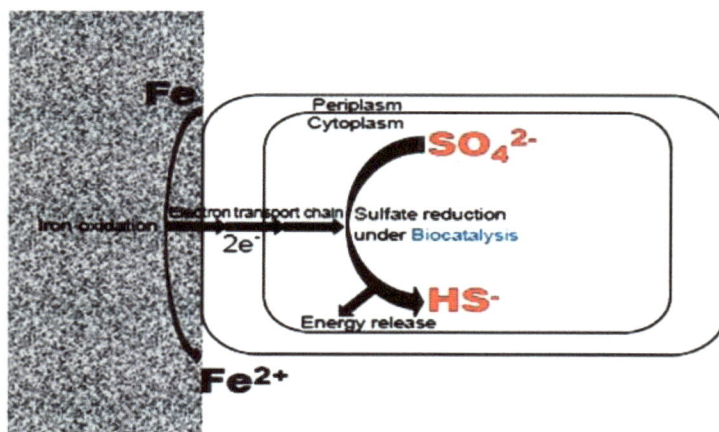

Fig. (3). Representation of consumption of electrons released from iron oxidation by SRB for sulfate reduction [87].

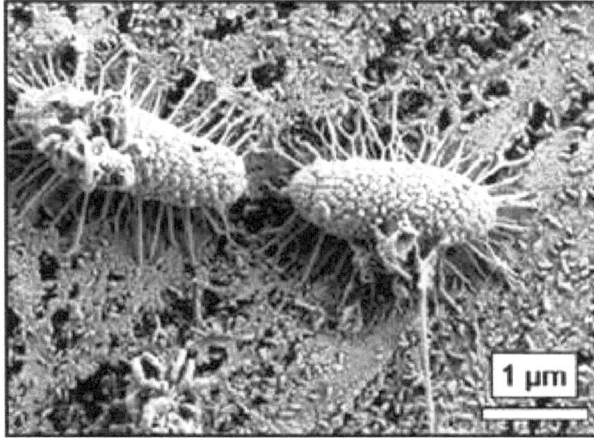

Fig. (4). SRB nanowires, which consume electrons from iron [89].

SRB also damages the metallic materials by following another mechanism, where the H_2S acts as poisons and inhibits the combination of hydrogen atoms from forming molecular hydrogen [82]. The atomic hydrogen is diffused into the metal matrix and combines therein with other hydrogen atoms to form molecular hydrogen. As a result of volume expansion of the crystal lattice due to the accumulated hydrogen molecules, internal stress generates and finally leads to the stress-induced cracking known as hydrogen embrittlement [83].

4.2. Biocatalytic Cathodic Sulfate Reduction (BCSR) Theory

Gu *et al.*, in 2009, proposed a new theory to explain MIC known as biocatalytic cathodic sulfate reduction theory (BCSR) [84]. It was assumed that SRB induced MIC occurs due to the dissolution of anodic material, *i.e.* Fe release electrons; and then the electrons move to the SRB cells and are consumed for sulfate reduction within the cytoplasm of the cells. This is known as the biocatalytic process. The corrosive SRB biofilm deposited on the iron surface mainly causes the MIC, and BCSR reaction follows due to biofilm bio-catalysis. The potential of the biological systems is taken at standard conditions Eo' like 25°C temperature, 1 M solute concentration, 1 partial bar pressure of the gas; however, the pH value is taken to be 7, not 0. The overall reactions are given as follows:

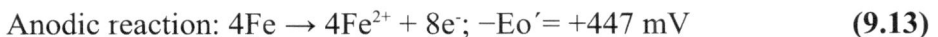

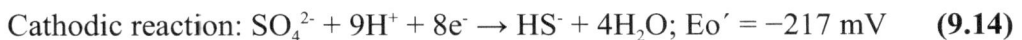

$$\text{Anodic reaction: } 4Fe \rightarrow 4Fe^{2+} + 8e^-; -Eo' = +447 \text{ mV} \qquad \textbf{(9.13)}$$

$$\text{Cathodic reaction: } SO_4^{2-} + 9H^+ + 8e^- \rightarrow HS^- + 4H_2O; Eo' = -217 \text{ mV} \qquad \textbf{(9.14)}$$

In BCSR theory, there is no physical cathode rather, SRB biofilm acts as the cathode, which is called biocathode [84]. This is because the sulfate reduction occurs inside SRB cells. The BCSR theory gives the idea of the overall effect as the sulfate reduction process is more complicated in the cytoplasm of the SRB cells than the reaction (9.13). Reaction (9.14) is very slow if there is no bio-catalysis from biofilms. The reaction occurs because of the bio-catalysis due to the hydrogenase enzyme system in hydrogenase-positive SRB cells using H_2 as an electron carrier [69]. The hydrogenase lowers the activation energy to speed up the reaction.

On the basis of bioenergetics, BCSR theory explains in a better way about how MIC occurs. For example, the MIC pitting corrosion of iron due to SRB biofilm can be explained by BCSR theory. The anaerobic respiration utilizing organic carbons, *e.g.*, from lactate, as the electron donor and sulfate as the electron acceptor, is due to planktonic SRB. The oxidation reaction showing how SRB interacts with the lactate was given by Thaeur *et al.* [85] as:

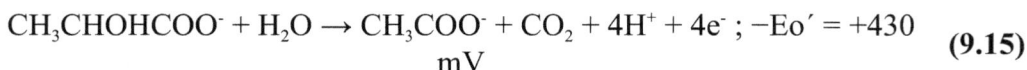

$$CH_3CHOHCOO^- + H_2O \rightarrow CH_3COO^- + CO_2 + 4H^+ + 4e^- ; -Eo' = +430 \\ mV \tag{9.15}$$

$$SO_4^{2-} + 2CH_3CHOHCOO^- \rightarrow 2CH_3COO^- + 2CO_2 + HS^- + OH^- + H_2O \tag{9.16}$$

$\Delta Go' = -164.4$ kJ/mol<0 (where $\Delta Eo' = +213$ mV)

Reaction (9.15) is the anaerobic respiration of SRB indicating the −ve Gibbs free energy, energy release in the reaction, and spontaneity. It was found that no reaction has occurred when the lactate and sulfate are mixed without SRB, *i.e.* without biocatalysis. So, there is the necessity of enzymes for lactate oxidation and sulfate reduction reactions. The redox reaction (9.16) takes place in the cytoplasm inside the SRB cells [85]. In this reaction, the planktonic SRB is not directly involved in the corrosion; rather the diffusion of organic carbon into the cytoplasm of the SRB as the electron donor plays the role. Hence, microorganisms are not directly involved to corrode the iron.

The driving force for the diffusion of organic carbon from the bulk to the surface of iron is the concentration gradient [86]. The driving force must be sufficient enough for the diffusion of organic carbon into the sessile SRB cells deep inside the biofilm, which is very dense on the iron surface. The energy consideration is again needed to satisfy the requirement of energy for their own metabolic activity, wherein the top layer of the biofilm consumes most of the organic carbon diffused into the biofilm. Thus, a situation is created, whenever the energy is large enough for the diffusion of organic carbon. But due to the consumption of the top layer of biofilm, there is no sufficient carbon to reach the sessile cells deeper inside the biofilm. Hence, maintenance energy is needed for the survival of sessile cells, which is possible by shifting from organic carbon oxidation to iron oxidation for the supply of electrons to reduce sulfate and produce energy. The energy produced by iron oxidation is found to be slightly more than lactate oxidation, and the electron released from iron oxidation is consumed by sulfate reduction with the support of bio-catalysis. When compared to lactate, iron is less soluble, and hence it cannot directly involve in the cytoplasm to donate electrons for the sulfate reduction. Hence, the electron transport chain provided by SRB to transport the extracellular electrons and thus leading to corrosion. Hernandez and Newman suggested that for producing the energy by some microorganisms, this transport phenomenon known as extracellular electron transfer is one of the most fundamental methods. The SRB consumption of electrons from iron for sulfate reduction is represented as in Fig. (**3**) [87]. Du *et al.* found the electrons are transported either by (a) direct electron transfer (DET), or (b) mediated electron transfer (MET) [88]. In DET the electrons are transported through the protein called pili or c type cytochrome also known as nanowires.

Sherar *et al.*, suggested the SRB produces nanowire to consume electron from iron in a modified Postgate medium with no organic carbon [89]. Transportation of electrons takes place from iron to SRB cells through the pili, and it was represented in Fig. (**4**), already by Scherar *et al.* [89].

In MET, the electrons transport *via* some specific soluble chemicals, which are electron mediators also known as shuttles, through their cell membrane will be taken place. In oil and gas fields, commonly the bacterium Shewanella putrefaciens is found which is existed with SRB in biofilm [90]. This bacterium secretes FAD, riboflavin electron mediators and these shuttles are utilized for the efficient transportation of electrons. Hence, it can make faster corrosion. The energy released due to corrosion is responsible for the co-existence of both SRB and Shewanella putrefaciens in the biofilm for which their synergistic effect can cause faster corrosion. In fact, the fast pitting requires the synergistic effect as the single strain of bacterium existing in the biofilm does not have an influence on this type of corrosion. Fig. (**5**) shows the effect of DET and MET in the transportation of electrons between electrodes and bacterial cells.

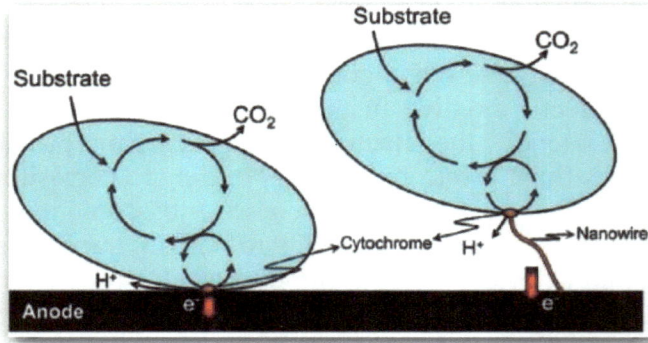

Fig. (5). Representation of DET and MET: A→C (A- direct contact of cytochrome, B-pili (nanowires) and C-mediators [91, 92].

Venzlaff *et al.*, proposed another mechanism based on electrochemical evidence of direct electron uptake released from carbon steel by sulfate-reducing bacteria [93]. Fig. (**6**) represents the role of electron transfer in MIC.

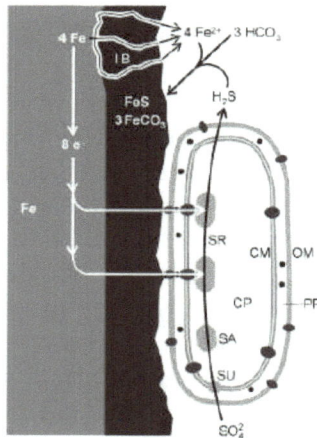

Fig. (6). Direct uptake of electrons released from iron by sulfate reducing bacteria responsible for corrosion [93].

Nitrate reducing bacteria (NRB) is another microbe causing MIC. Sometimes nitrate is used in oil reservoirs to increase NRB growth to minimize SRB growth. Hence, to reduce the biogenic H_2S production, causing pipeline corrosion and reservoir souring [94, 95]. Iron oxidation and nitrate reduction are thermodynamically favorable [96]. Pseudomonas aeruginosa and Bacillus licheni grow as nitrate-reducing bacteria which are electron acceptors in anaerobic condition for their respiration. Pseudomonas aeruginosa biofilm grown an aerobically on 304 SS caused pitting corrosion and Bacillus licheni form is biofilm grown on carbon steel caused corrosion more than caused by only the sulfate-reducing bacteria [96, 97]. Here, it is suggested that the nitrate injection in souring mitigation should not leave the surplus nitrate. Agricultural runoff carries nitrate contaminants to the soil, which is again responsible for the MIC in soil [98]. BCSR theory can also explain the MIC caused by nitrate reduction and can be modified as biocatalytic cathodic nitrate reduction (BCNR). The cathodic reduction reactions follow as:

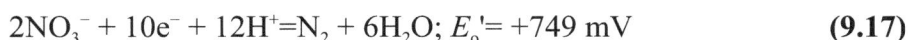

$$2NO_3^- + 10e^- + 12H^+ = N_2 + 6H_2O; \; E_o' = +749 \text{ mV} \tag{9.17}$$

$$NO_3^- + 8e^- + 10H^+ = NH_4^+ + 3H_2O; \; E_o' = +358 \text{ mV} \tag{9.18}$$

BCSR and BCNR theories can explain MIC caused by SRB and NRB corrosive bacteria. These are known as 'XRB' where X means oxidants, *e.g.*, sulfate, nitrate or nitrite or carbon dioxide; and B represents bugs, *e.g.*, archaea, bacteria, and eukaryotes [99].

4.3. Acid Producing Bacteria (APB)

It has already been found that the MIC can be caused by organic acids present under the APB biofilms with pH lower than the bulk fluid [100, 101]. Oxidation of iron and the associated proton reduction has been found to be thermodynamically favorable, and the reaction goes without biocatalyst is unlike sulfate or nitrate reduction [102]. These acids are more corrosive than the strong acids at the same pH because the organic acids are wearing acids and the undissociated organic acids serve as a reservoir that releases more protons [103]. It is known that the lower pH of the environment is detrimental to the passivation of the metal [104].

4.4. Archaea

Archea like bacteria has no membrane-bound nucleus found abundantly in the environment [105, 106]. Some of them are sulfate or nitrate reducers, and some are methanogens [107 - 109]. Some of them can live even in an environment of high temperature and pressure [110]. Thermophilic sulfate-reducing archea (SRA) is found abundantly in marine hydrothermal systems [111, 112]. Like sulfate-reducing bacteria, the sulfate-reducing archea also uptake the electrons during their respiration and corrosive in nature.

5. MITIGATION OF MIC/BIOFILMS

5.1. Conventional Mitigation Methods

It is understood that biofilms are mainly responsible for MIC. Hence, mitigation of biofilms would be a necessary condition to mitigate the MIC. Mechanical cleaning and biocides are conventional methods for the mitigation of MIC in industrial sectors [113]. Pigging is one of the important methods to prevent or minimize MIC [114]. Pigging is a commonly used method in oil and gas industries, pipelines, where it scrubs off the biofilm and the corrosion products [115]. Pigging coupled with a biocide solution, has been commonly used. Slower the pigging time longer will be the exposure of the biocides. The use of pigs is not used for all types of pipelines, *e.g.*, restrictive to small diameter pipelines and bends and connects as pigs are designed for large diameter pipelines.

On the other hand, uses of biocides are most common to mitigate MIC. THPS and glutaraldehyde are common biocides widely used because they are biodegradable and highly efficient against microorganisms. THPS breaks the disulfide bonds in proteins in microbes. The degraded THPS converts to trishydroxymethyl phosphine oxide, which is no more a biocide [116]. Glutaraldehyde, a cross-linking agent, undergoes cross-link amino acids in cell walls [117]. Sessile cells present in the biofilms are 10 to 1000 times more resistant to antimicrobial agents than planktonic cells and protected from the biofilm. Hence, a high biocide concentration is required to treat them [118]. The use of a particular biocide can decrease the concentration of comparatively weak cells and thus increase the concentration of strongly resisting cells. Hence, the treatment may require a high concentration of biocides. The use of a high concentration of biocides has side effects of raising economic and environmental issues.

5.2. Biocide Enhancers

The treatment of sessile cells has a strong resistance to antimicrobial agents more than planktonic cells, and it requires a high dosage of biocides [119]. As the treatment is not sufficient to eliminate the microbes completely in a cycle, the biofilm may again be formed. So, the treatment has to be repeated, and in doing so, the biocides with more resistant cells may have resulted. Thus, a higher dose of biocides will be demanded, which may cause economic and environmental issues. The Tetrakis (hydroxymethyl) phosphonium sulfate (THPS) and glutaraldehyde are the most widely used biocides. So, these biocides are to be enhanced to reduce the biocide dosage. Biocide enhancers are not the biocides, but they can enhance the biocides through different mechanisms and thus reduce the dosage of biocides or make them more efficient for a given concentration. Ethylenediaminetetraacetic acid (EDTA) and ethylenediaminedisuccinate (EDDS) is known to be biocide enhancers to mitigate biofilms [120, 121]. These are the biodegradable biocide enhancers. Na-EDTA used to eliminate the acidity of the acid form, and they accumulate in the water bodies as it is slow biodegradable, whereas EDDS replaced EDTA because of its readily biodegradable property. It was found that EDDS can enhance THPS and glutaraldehyde against planktonic cells of *D. Vulgaris* and *Desulfovibrio alaskensis* and the glutaraldehyde in SRB biofilm in the prevention and removal test [122, 123]. EDDS can easily be consumed by various ions in a liquid, so its high dosage is required for the large-scale treatment and hence it is not recommended.

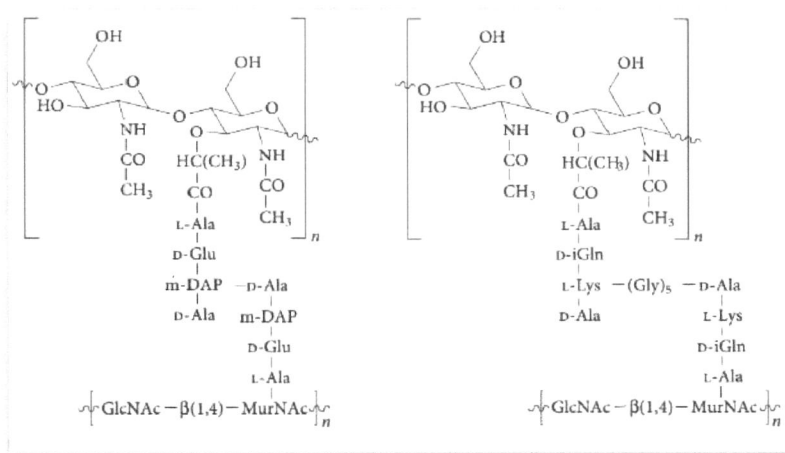

Fig. (7). Representation of two types of peptidoglycan→ **(A)** DAP type; and **(B)** Lys-type [130].

D-amino acids are known to be a natural chemical found in plants, animals, and humans [124]. Various D-amino acids like D-tyrosine (D-tyr), D-methionine (D-met), D-leucinet (D-leu), and D-tryptophan (D-trp) had shown the disassembly of Bacillus subtilis, Staphylococcus aureus, and P. aeruginos a biofilms [125]. However, the mechanism of how the D-amino acids disperse biofilms is not yet understood. Several workers proposed the mechanism for MIC. Cava *et al.* [126] proposed the role of D-amino acids in remodelling the cell wall structure. Lam *et al.* proposed that the synthesis of peptidoglycan a strong polymer acts as a stress-bearing component that can be down-regulated by D-amino acids [127]. Some other workers proposed the mechanism in different ways [128 - 130].

Kolodkin-Gal *et al.* [125] and Yu *et al.* [131] suggested that the biofilms dissembled by D-amino acids without inhibiting the bacterial growth. It was suggested that the D-amino acids could work as signal molecules to disperse biofilms. D-alanine (D-ala) terminus being replaced by D-amino acids in the cell wall results in the biofilm disassembling (Fig. **7**). This is a naturally occurring process in the peptide side chains of peptidoglycan of bacterial cell walls. D-amino acids make the bacteria to control the cell wall reaction to sustain in the modified environment, modification of cell wall assembly, and biofilm attachment.

CONCLUSION

Microbiologically influenced corrosion is based on the various electron consumption processes by microorganisms like sulfate-reducing bacteria, nitrate-reducing bacteria, acid-producing bacteria, archea, *etc*. Several workers proposed the mechanisms for the MIC processes. The most accepted theories are Cathodic depolarization theory, Biocatalytic cathodic sulfate reduction (BCSR), biocatalytic cathodic nitrate reduction (BCNR) however, more work is required to understand clearly the MIC mechanism. How electron transfer from an iron surface to the cytoplasm is yet to be clear to understand MIC. How the electron mediators shuttle the electrons into or out of the cell membrane needs further research concentration.Mitigation of MIC is based on the mitigation of the biofilm formation. Mechanical cleaning and biocides are popular conventional methods to mitigate MIC. The most widely adopted biocides are Tetrakis (hydroxymethyl) phosphonium sulfate (THPS) and glutaraldehyde. However, these biocides are not applicable in all cases as it may require in some cases a high dosage and create environmental and economic issues. Biocide enhancers, on the other hand, improve the efficiency to mitigate MIC without environmental and economic issues. Ethylenediaminetetraacetic acid (EDTA) and ethylenediaminedisuccinate (EDDS) are commonly used biocide enhancers. But their presence in high dosage may have side effects to cause environmental issues. D-amino acids are known to be efficient biocide enhancers and are most widely used. The effects on mitigation of different field biofilm by combined D-amino acids need the research concentration.

CONSENT FOR PUBLICATION

Not applicable.

CONFLICT OF INTEREST

The author declares no conflict of interest, financial or otherwise.

ACKNOWLEDGEMENTS

Declared none.

REFERENCES

[1]   Axelsen, S.B.; Rogne, T. Do micro-organisms "eat" metal?" Microbiologically influenced corrosion of industrial materials. *Contract No. BRRT-CT98-5084,* **1998**.

[2]   Videla, H.A. *Manual of Biocorrosion,* 1st ed; CRC-Press, **1996**, pp. 13-45.

[3]   Flemming, H-C. Biofouling, and microbiologically influenced corrosion (MIC) An economic and technical overview, in Heitz, E., Sand W., and Flemming. H.-C. (eds.), Microbial Deterioration of Materials. Springer-Verlag, Berlin-New York, 1996, 5–14.

[4]   Videla, H.A. Prevention and control of biocorrosion. *Int. Biodeterior. Biodegradation,* **2002**, *49*, 259-270.
[http://dx.doi.org/10.1016/S0964-8305(02)00053-7]

[5]   Watkinson, R.J. *Hydrocarbon Degradation in Microbial Problems and Corrosion in Oil and Oil Products Storage*; The Institute of Petroleum: London, UK, **1984**.

[6]   Gaylarde, C.C.; Bento, F.M. Microbial contamination of stored hydrocarbon fuels and its control. *Rev. Microbiol.,* **1999**, *30*, 1-10.
[http://dx.doi.org/10.1590/S0001-37141999000100001]

[7]   Passman, F.J., Ed. *Fuel and fuel system microbiology, fundamentals, diagnosis, and contamination control*; ASTM International, **2003**.
[http://dx.doi.org/10.1520/MNL47-EB]

[8]   Videla, H.A.; Herrera, L.K. Microbiologically influenced corrosion: looking to the future. *Int. Microbiol.,* **2005**, *8*(3), 169-180.
[PMID: 16200495]

[9]   Dodos, G.S.; Perdiou, V.; Zannikos, F. Effect of biodiesel in the microbiological growth through the diesel fuel supply chain. *Proceedings from the 7th Pan-Hellenic Scientific Conference on Chemical Engineering,* **2009**.

[10]   Schleicher, T.; Werkmeister, R.; Russ, W.; Meyer-Pittroff, R. Microbiological stability of biodiesel-diesel-mixtures. *Bioresour. Technol.,* **2009**, *100*(2), 724-730.
[http://dx.doi.org/10.1016/j.biortech.2008.07.029] [PMID: 18793842]

[11]   Sørensen, G.; Pedersen, D.V.; Nørgaard, A.K.; Sørensen, K.B.; Nygaard, S.D. Microbial growth studies in biodiesel blends. *Bioresour. Technol.,* **2011**, *102*(8), 5259-5264.
[http://dx.doi.org/10.1016/j.biortech.2011.02.017] [PMID: 21376581]

[12]   Al-Shamari, A.R.; Al-Mithin, A.W.; Prakash, S.; Islam, M.; Biedermann, A.J.; Methew, A. Some empirical observation about bacteria proliferation and corrosion damage morphology in Kuwait oilfield waters , **2013**.

[13]   Implants, B. Corrosion and its Prevention: A Review. *Recent Pat. Corros. Sci.,* *2*(1), 40-54.

[14]   Manam, N.S.; Harun, W.S.W.; Shri, D.N.A.; Ghani, S.A.C.; Kurniawan, T.; Ismail, M.H.; Ibrahim, M.H. Study of corrosion in biocompatible metals for implants: A review. *J. Alloys Compd.,* **2017**, *701*, 698-715.
[http://dx.doi.org/10.1016/j.jallcom.2017.01.196]

[15]   Manivasagam, G.; Dhinasekaran, D.; Rajamanickam, A. Biomedical Implants:Corrosion and its Prevention: A Review. *Recent Pat. Corros. Sci.,* **2010**, *2*(1), 40-54.
[http://dx.doi.org/10.2174/1877610801002010040]

[16]   Chandrasatheesh, C.; Jayapriya, J.; George, R.P. *Eng. Fail. Anal.,* **2014**, *42*, 133-142.

[http://dx.doi.org/10.1016/j.engfailanal.2014.04.002]

[17]   Li, H.; Zhou, E.; Ren, Y.; Zhang, D.; Xu, D.; Yang, C. Investigation of microbiologically influenced corrosion of high nitrogen nickel-free stainless steel by Pseudomonas aeruginosa. *Corros. Sci.,* **2016**, *111*, 811-821.
[http://dx.doi.org/10.1016/j.corsci.2016.06.017]

[18]   Garrett, J.H. *The Action of water on lead*; Lewis, London, **1981**.

[19]   Gaines, R.H. Bacterial Activity as a Corrosive Influence in the Soil. *Ind. Eng. Chem.,* **1910**, *2*, 128-130.
[http://dx.doi.org/10.1021/ie50016a003]

[20]   Ellis, D. *Iron Bacteria*; Methuden: London, **1919**.

[21]   Von WolzogenKuhr. C.A.H. Van der Vlugt, L.R., Degrafiteerign van Giebijzeralselectrochemisch process in anaerobe gronden, Water (den Haag) 18, 147. *Translated in Corrosion,* **1934**, *17*, 293-299.

[22]   Starkey, R.L. Transformation of sulphur by microorganisms. *Ind. Eng. Chem.,* **1956**, *48*, 1429-1437.
[http://dx.doi.org/10.1021/ie51400a022]

[23]   Hadley, R.F. The influence of Sporovibriodesulfuricans on the Current and Potential Behaviour of Corroding Iron National Bureau of Standards Corrosion Conference, 1943.

[24]   Duchon, K.; Miller, L.B. The effect of Chemical Agents on Iron Bacteria. *Paper Trade J.,* **1948**, *124*, 47.

[25]   Butlin, K.R.; Adams, M.E.; Thomas, M. The isolation and cultivation of sulphate-reducing bacteria. *J. Gen. Microbiol.,* **1949**, *3*(1), 46-59. a
[PMID: 18126507]

[26]   Butlin, K.R.; Adams, M.E.; Thomas, K. Internal Corrosion of Ferrous Pipes conveying water. *Nature,* **1949**, *163*, 26. b
[http://dx.doi.org/10.1038/163026a0] [PMID: 18106144]

[27]   Wanklyn, J.N.; Spruit, C.J.P. Influence of SRB on the Corrosion Potential of Iron. *Nature,* **1952**, *1952*(169), 928-929.
[http://dx.doi.org/10.1038/169928b0]

[28]   Skybalski, W.; Olsen, F. Aerobic Microbial Corrosion of Water Pipes. *Clin. Orthop. Relat. Res.,* **1949**, *1949*(6), 405-414.

[29]   Uhlig, H.H., Ed. *Corrosion Handbook,* 4th ed; John Wiley: N.Y, **1953**.

[30]   Kulman, E.E. Microbiological corrosion of buried steel pipe. *Corros. Sci.,* **1953**, *9*, 11-18.
[http://dx.doi.org/10.5006/0010-9312-9.1.11]

[31]   Booth, G.H.; Tiller, A.K. Polarisation studies of mild steel in cultures of sulfate reducing bacteria. *Trans. Faraday Soc.,* **1960**, *58*, 2510-2516.
[http://dx.doi.org/10.1039/TF9625802510]

[32]   Booth, G.H.; Tiller, A.K. Polarization studies of mild steel in cultures of sulphate-reducing bacteria. *Trans. Faraday Soc.,* **1960**, *56*, 1689-1696.
[http://dx.doi.org/10.1039/tf9605601689]

[33]   Booth, G.H.; Tiller, A.K. Cathodic characteristics of mild steel in suspensions of sulphate reducing bacteria. *Corros. Sci.,* **1968**, *8*, 583-600.
[http://dx.doi.org/10.1016/S0010-938X(68)80094-0]

[34]   Horvath, J. Electrochemical studies on the corrosion of steel by microbiological oxidation of reduced inorganic compounds. In: *Proc. First Int. Cong. Metal Corr*; London, **1962**; pp. 345-351.

[35]   Iverson, W.P. Direct evidence for the cathodic depolarization theory of bacterial corrosion. *Science,* **1966**, *151*(3713), 986-988.
[http://dx.doi.org/10.1126/science.151.3713.986] [PMID: 17796779]

[36]   Lee, W.; Lewandowski, Z.; Nielsen, P.H.; Hamilton, W.A. Role of sulfate-reducing bacteria in corrosion of mild steel: a review. *Biofouling*, **1995**, *8*, 165-194.
[http://dx.doi.org/10.1080/08927019509378271]

[37]   Pankhania, I.P. Hydrogen metabolism in sulphate-reducing bacteria and its role in anaerobic corrosion. *Biofouling*, **1988**, *1*, 27-47.
[http://dx.doi.org/10.1080/08927018809378094]

[38]   Iverson, W.P. Corrosion of iron and formation of iron phosphide by Desulfovibrio desulfuricans. *Nature*, **1968**, *217*(5135), 1265-1267.
[http://dx.doi.org/10.1038/2171265a0] [PMID: 4868375]

[39]   Ferris, F.G.; Schultze, S.; Witten, T.C.; Fyfe, W.S.; Beveridge, T.J. Metal interactions with microbial biofilms in acidic and neutral pH environments. *Appl. Environ. Microbiol.*, **1989**, *55*(5), 1249-1257.
[http://dx.doi.org/10.1128/AEM.55.5.1249-1257.1989] [PMID: 16347914]

[40]   Javaherdashti, R. A review of some characteristics of MIC caused by sulfate-reducing bacteria: Past, present and future. *Anti-Corros. Methods Mater.*, **1999**, *46*, 173-180.
[http://dx.doi.org/10.1108/00035599910273142]

[41]   Mansfeld, F.; Little, B. A technical review of electrochemical techniques applied to microbiologically influenced corrosion. *Corros. Sci.*, **1991**, *32*, 247-272.
[http://dx.doi.org/10.1016/0010-938X(91)90072-W]

[42]   Videla, H.A.; Characklis, W.G. Biofouling and microbially influenced corrosion. *Int. Biodeterior. Biodegradation*, **1992**, *29*, 195-212.
[http://dx.doi.org/10.1016/0964-8305(92)90044-O]

[43]   Walsh, D.; Pope, D.; Danford, M.; Huff, T. The effect of microstructure on microbiologically influenced corrosion. *JOM J. Miner. Met. Mater. Soc.*, **1993**, *45*, 22-30.
[http://dx.doi.org/10.1007/BF03222429]

[44]   Magot, M. Indigenous microbial communities in oil fields. In: *Petroleum Microbiology*; Oliver, B.; Magot, B., Eds.; American Society of Microbiology: Washington, DC, USA, **2005**; pp. 21-34.

[45]   Larsen, J.; Rasmussen, K.; Pedersen, H.; Sørensen, K.; Lundgaard, T.; Skovhus, T.L. Consortia of MIC bacteria and archaea causing pitting corrosion in top side oil production facilities. *Proceedings of the Corrosion 2010*, San Antonio, TX, USA**2010**, pp. 14-18.

[46]   Vigneron, A.; Alsop, E.B.; Chambers, B.; Lomans, B.P.; Head, I.M.; Tsesmetzis, N. Complementary microorganisms in highly corrosive biofilms froman offshore oil production facility. *Appl. Environ. Microbiol.*, **2016**, *82*(8), 2545-2554.
[http://dx.doi.org/10.1128/AEM.03842-15] [PMID: 26896143]

[47]   Usher, K.M.; Kaksonen, A.H.; Mac Leod, L.D. Marine rust tubercles harbour iron corroding archaea and sulphate reducing bacteria. *Corros. Sci.*, **2014**, *83*, 189-197.
[http://dx.doi.org/10.1016/j.corsci.2014.02.014]

[48]   Boopathy, R.; Daniels, L. Effect of pH on anaerobic mild steel corrosion by methanogenic bacteria. *Appl. Environ. Microbiol.*, **1991**, *57*(7), 2104-2108.
[http://dx.doi.org/10.1128/AEM.57.7.2104-2108.1991] [PMID: 16348530]

[49]   Videla, H.A.; Herrera, L.K. Understanding microbial inhibition of corrosion. A comprehensive overview. *Int. Biodeterior. Biodegradation*, **2009**, *63*, 896-900.
[http://dx.doi.org/10.1016/j.ibiod.2009.02.002]

[50]   Herrera, L.K.; Videla, H.A. Role of iron-reducing bacteria in corrosion and protection of carbon steel. *Int. Biodeterior. Biodegradation*, **2009**, *63*, 891-895.
[http://dx.doi.org/10.1016/j.ibiod.2009.06.003]

[51]   Heukelekian, H.; Heller, A. Relation between food concentration and surface for bacterial growth. *J. Bacteriol.*, **1940**, *40*(4), 547-558.

[http://dx.doi.org/10.1128/JB.40.4.547-558.1940] [PMID: 16560368]

[52]   Telegdi, Judit; Shaban, Abdul; Laszlo, L. Trends c0008 Microbiologically influenced corrosion (MIC), Trends in Oil and Gas Corrosion Research and Technologies. *Elsevier publishing.,* [http://dx.doi.org/10.1016/B978-0-08-101105-8.00008-5]

[53]   Jefferson, K.K. What drives bacteria to produce a biofilm? *FEMS Microbiol. Lett.,* **2004**, *236*(2), 163-173.
[http://dx.doi.org/10.1111/j.1574-6968.2004.tb09643.x] [PMID: 15251193]

[54]   Flemming, H.C.; Wingender, J. The biofilm matrix. *Nat. Rev. Microbiol.,* **2010**, *8*(9), 623-633.
[http://dx.doi.org/10.1038/nrmicro2415] [PMID: 20676145]

[55]   Tiller, A.K. Aspect ofmicrobial corrosion. In: *Corrosion Processes*; Parkins, R.N., Ed.; Applied Science Publication: London, UK, **1982**; pp. 115-159.

[56]   Brenda, L.; Patricia, W. An overview of microbiologically influenced corrosion of metals and alloys used in the storage of nuclear wastes. *Can. J. Microbiol.,* **1996**, *42*, 367-374.
[http://dx.doi.org/10.1139/m96-052]

[57]   Little, B.; Wagner, P.; Mansfeld, F. Microbiologically influenced corrosion of metals and alloys. *Int. Mater. Rev.,* **1991**, *36*, 253-272.
[http://dx.doi.org/10.1179/imr.1991.36.1.253]

[58]   Characklis, W. Bioengineering report: Fouling biofilmdevelopment: A process analysis. *Biotechnol. Bioeng.,* **1981**, *23*, 1923-1960.
[http://dx.doi.org/10.1002/bit.260230902]

[59]   Pedersen, A.; Hermansson, M. The effects on metal corrosion by Serratiamarcescens and a Pseudomonas SP. *Biofouling,* **2009**, 313-322.

[60]   Callow, M.E.; Fletcher, R.L. The influence of low surface energy materials on bioadhesion—A review. *Int. Biodeterior. Biodegradation,* **1994**, *34*, 333-334.
[http://dx.doi.org/10.1016/0964-8305(94)90092-2]

[61]   Blackwood, D.J.; Seah, K.W.H.; Teoh, S.H. Corrosion of metallic implants. , **2004**.
[http://dx.doi.org/10.1142/9789812562227_0003]

[62]   Little, B.J.; Lee, J.S. Microbiologically influenced corrosion: An update. *J. Int. Mater. Rev.,* **2014**, *59*, 384-393.
[http://dx.doi.org/10.1179/1743280414Y.0000000035]

[63]   Videla, H.A. *Manual of Biocorrosion,* 1st ed; CRC-Press, **1996**, pp. 13-45.

[64]   Iverson, W.P. Direct Evidence for the Cathodic Depolarization Theory of Bacterial Corrosion; Defense Technical Information Center: Fort Belvoir, VA, USA *Report No. DTIC AD0476409,* **1965**.

[65]   Enning, D.; Garrelfs, J. Corrosion of iron by sulfate-reducing bacteria: new views of an old problem. *Appl. Environ. Microbiol.,* **2014**, *80*(4), 1226-1236.
[http://dx.doi.org/10.1128/AEM.02848-13] [PMID: 24317078]

[66]   Araujo, J.C.; Téran, F.C.; Oliveira, R.A.; Nour, E.A.A.; Montenegro, M.A.P.; Campos, J.R.; Vazoller, R.F. Comparison of hexamethyldisilazane and critical point drying treatments for SEM analysis of anaerobic biofilms and granular sludge. *J. Electron Microsc. (Tokyo),* **2003**, *52*(4), 429-433.
[http://dx.doi.org/10.1093/jmicro/52.4.429] [PMID: 14599106]

[67]   Cord-Ruwisch, R. Microbially influenced corrosion of steel. In: *Environmental microbe-metal interactions*; Lovley, D.R., Ed.; ASM Press: Washington, DC, **2000**; pp. 159-173.

[68]   Appia-Ayme, C.; Guiliani, N.; Ratouchniak, J.; Bonnefoy, V. Characterization of an operon encoding two c-type cytochromes, an aa(3)-type cytochrome oxidase, and rusticyanin in Thiobacillus ferrooxidans ATCC 33020. *Appl. Environ. Microbiol.,* **1999**, *65*(11), 4781-4787.
[http://dx.doi.org/10.1128/AEM.65.11.4781-4787.1999] [PMID: 10543786]

[69]   Gu, T.; Xu, D. Why are some microbes corrosive and some not? in: Corrosion/2013 Paper No. C2013-0002336 *NACE International, Houston, TX,* **2013**.

[70]   Little, B.J.; Ray, R.I.; Pope, R.K. Relationship between corrosion and the biological sulfur cycle: a review. *Corrosion,* **2000**, *56*, 433-443.
[http://dx.doi.org/10.5006/1.3280548]

[71]   Thierry, D.; Sand, W. Microbially Influenced Corrosion. In: *Corrosion Mechanisms in Theory and Practice*; Marcus, P., Ed.; CRC Press, **2002**; pp. 583-603.
[http://dx.doi.org/10.1201/9780203909188.ch16]

[72]   Gu, T.; Xu, D. Why are some microbes corrosive and some not? in: Corrosion/2013 Paper No. C2013-0002336. NACE International, Houston, TX.

[73]   Widdel, F. Microbial corrosion. In: *Biotechnology focus 3*; Finn, R.K.; Prave, P.; Schlingmann, M.; Crueger, W.; Esser, K.; Thauer, R.; Wagner, F., Eds.; Hanser: Munich, Germany, **1992**; pp. 277-318.

[74]   Venzlaff, H.; Enning, D.; Srinivasan, J.; Mayrhofer, K.; Hassel, A.W.; Widdel, F.; Stratmann, M. Accelerated cathodic reaction in microbial corrosion of iron due to direct electron uptake by sulfate-reducing bacteria. *Corros. Sci.,* **2012**, *66*, 88-96.
[http://dx.doi.org/10.1016/j.corsci.2012.09.006]

[75]   Dinh, H.T.; Kuever, J.; Mussmann, M.; Hassel, A.W.; Stratmann, M.; Widdel, F. Iron corrosion by novel anaerobic microorganisms. *Nature,* **2004**, *427*(6977), 829-832.
[http://dx.doi.org/10.1038/nature02321] [PMID: 14985759]

[76]   Mori, K.; Tsurumaru, H.; Harayama, S. Iron corrosion activity of anaerobic hydrogen-consuming microorganisms isolated from oil facilities. *J. Biosci. Bioeng.,* **2010**, *110*(4), 426-430.
[http://dx.doi.org/10.1016/j.jbiosc.2010.04.012] [PMID: 20547365]

[77]   Enning, D.; Garrelfs, J. Corrosion of iron by sulfate-reducing bacteria: new views of an old problem. *Appl. Environ. Microbiol.,* **2014**, *80*(4), 1226-1236.
[http://dx.doi.org/10.1128/AEM.02848-13] [PMID: 24317078]

[78]   Enning, D.; Venzlaff, H.; Garrelfs, J.; Dinh, H.T.; Meyer, V.; Mayrhofer, K.; Hassel, A.W.; Stratmann, M.; Widdel, F. Marine sulfate-reducing bacteria cause serious corrosion of iron under electroconductive biogenic mineral crust. *Environ. Microbiol.,* **2012**, *14*(7), 1772-1787.
[http://dx.doi.org/10.1111/j.1462-2920.2012.02778.x] [PMID: 22616633]

[79]   Pereira, I.A.; Ramos, A.R.; Grein, F.; Marques, M.C.; da Silva, S.M.; Venceslau, S.S. A comparative genomic analysis of energy metabolism in sulfate reducing bacteria and archaea. *Front. Microbiol.,* **2011**, *2*, 69.
[http://dx.doi.org/10.3389/fmicb.2011.00069] [PMID: 21747791]

[80]   Rabus, R.; Hansen, T.; Widdel, F. Dissimilatory sulfate- and sulfurreducing prokaryotes. In: *The prokaryotes*; Dworkin, M.; Schleifer, K-H.; Stackebrandt, E., Eds.; Springer: New York, NY, **2006**; pp. 659-768.
[http://dx.doi.org/10.1007/0-387-30742-7_22]

[81]   Elboujdaini, M. Hydrogen-induced cracking and sulfide stress cracking. In: *Uhlig's corrosion handbook,* 3rd ed; Revie, R.W., Ed.; Wiley: Hoboken, NJ, **2011**; pp. 183-194.
[http://dx.doi.org/10.1002/9780470872864.ch15]

[82]   Radkevych, O.I.; Pokhmurs'kyi, V.I. Influence of hydrogen sulfide on serviceability of materials of gas field equipment. *Mater. Sci.,* **2001**, *37*, 319-332.
[http://dx.doi.org/10.1023/A:1013275129001]

[83]   Gu, T.; Zhao, K.; Nesic, S. A practical mechanistic model for MIC based on a biocatalyticcathodic sulfate reduction (BCSR) theory, in: Corrosion/2009 Paper No. 09390. NACE International, Houston, TX..

[84]   Rosenbaum, M.; Aulenta, F.; Villano, M.; Angenent, L.T. Cathodes as electron donors for microbial

metabolism: which extracellular electron transfer mechanisms are involved? *Bioresour. Technol.,* **2011**, *102*(1), 324-333.
[http://dx.doi.org/10.1016/j.biortech.2010.07.008] [PMID: 20688515]

[85]   Thauer, R.K.; Stackebrandt, E.; Hamilton, W.A. Energy metabolism Phylogeneticdiversity of sulphate-reducing bacteria , **2007**.

[86]   Gu, T.; Xu, D. Demystifying MIC mechanisms, in: Corrosion/2010 Paper No.10213. NACE International, Houston, TX..

[87]   DakeXu, Ph.D. thesis submitted in, Ohio University, 2013. 32.

[88]   Pant, D.; Van Bogaert, G.; Diels, L.; Vanbroekhoven, K. A review of the substrates used in microbial fuel cells (MFCs) for sustainable energy production. *Bioresour. Technol.,* **2010**, *101*(6), 1533-1543.
[http://dx.doi.org/10.1016/j.biortech.2009.10.017] [PMID: 19892549]

[89]   Sherar, B.W.A.; Power, I.M.; Keech, P.G.; Mitlin, S.; Southam, G.; Shoesmith, D.W. Characterizing the effect of carbon steel exposure in sulfide containing solutions to microbially induced corrosion. *Corros. Sci.,* **2011**, *53*, 955-960.
[http://dx.doi.org/10.1016/j.corsci.2010.11.027]

[90]   Martín-Gil, J.; Ramos-Sánchez, M.C.; Martín-Gil, F.J. Shewanella putrefaciens in a fuel-in-water emulsion from the Prestige oil spill. *Antonie van Leeuwenhoek,* **2004**, *86*(3), 283-285.
[http://dx.doi.org/10.1023/B:ANTO.0000047939.49597.eb] [PMID: 15539931]

[91]   Du, Z.; Li, H.; Gu, T. A state of the art review on microbial fuel cells: A promising technology for wastewater treatment and bioenergy. *Biotechnol. Adv.,* **2007**, *25*(5), 464-482.
[http://dx.doi.org/10.1016/j.biotechadv.2007.05.004] [PMID: 17582720]

[92]   Schröder, U. Anodic electron transfer mechanisms in microbial fuel cells and their energy efficiency. *Phys. Chem. Chem. Phys.,* **2007**, *9*(21), 2619-2629.
[http://dx.doi.org/10.1039/B703627M] [PMID: 17627307]

[93]   Venzlaff, H.; Enning, D.; Srinivasan, J.; Mayrhofer, K.J.J.; Hassel, A.W.; Widdel, F.; Stratmann, M. Accelerated cathodic reaction in microbial corrosion of iron due to direct electron uptake by sulfate-reducing bacteria. **2013**.
[http://dx.doi.org/10.1016/j.corsci.2012.09.006]

[94]   Fida, T.T.; Chen, C.; Okpala, G.; Voordouw, G. Implications of limited thermophilicity of nitrite reduction for control of sulfide production in oil reservoirs. *Appl. Environ. Microbiol.,* **2016**, *82*(14), 4190-4199.
[http://dx.doi.org/10.1128/AEM.00599-16] [PMID: 27208132]

[95]   Hasegawa, R.; Toyama, K.; Miyanaga, K.; Tanji, Y. Identification of crude-oil components and microorganisms that cause souring under anaerobic conditions. *Appl. Microbiol. Biotechnol.,* **2014**, *98*(4), 1853-1861.
[http://dx.doi.org/10.1007/s00253-013-5107-3] [PMID: 23912114]

[96]   Xu, D.; Li, Y.; Song, F.; Gu, T. Laboratory investigation of microbiologically influenced corrosion of C1018 carbon steel by nitrate reducing bacterium Bacillus licheniformis. *Corros. Sci.,* **2013**, *77*, 385-390.
[http://dx.doi.org/10.1016/j.corsci.2013.07.044]

[97]   Jia, R.; Yang, D.; Xu, D.; Gu, T. Anaerobic corrosion of 304 stainless steel caused by the Pseudomonas aeruginosa biofilm. *Front. Microbiol.,* **2017**, *8*, 2335.
[http://dx.doi.org/10.3389/fmicb.2017.02335] [PMID: 29230206]

[98]   Wan, H.; Song, D.; Zhang, D.; Du, C.; Xu, D.; Liu, Z.; Ding, D.; Li, X. Corrosion effect of Bacillus cereus on X80 pipeline steel in a Beijing soil environment. *Bioelectrochemistry,* **2018**, *121*, 18-26.
[http://dx.doi.org/10.1016/j.bioelechem.2017.12.011] [PMID: 29329018]

[99]   Gu, T.; Xu, D. *Why Are Some Microbes Corrosive and Some Not? Corrosion/2013 Paper No. 2013-2336*; NACE International: Orlando, Florida, **2013**.

[100] Xu, D.; Li, Y.; Gu, T. Mechanistic modeling of biocorrosion caused by biofilms of sulfate reducing bacteria and acid producing bacteria. *Bioelectrochemistry,* **2016**, *110*, 52-58.
[http://dx.doi.org/10.1016/j.bioelechem.2016.03.003] [PMID: 27071053]

[101] Kato, S. Microbial extracellular electron transfer and its relevance to iron corrosion. *Microb. Biotechnol.,* **2016**, *9*(2), 141-148.
[http://dx.doi.org/10.1111/1751-7915.12340] [PMID: 26863985]

[102] Gu, T. Theoretical modeling of the possibility of acid producing bacteria causing fast pitting biocorrosion. *J. Microb. Biochem. Technol.,* **2014**, *6*, 68-74.
[http://dx.doi.org/10.4172/1948-5948.1000124]

[103] Kryachko, Y.; Hemmingsen, S.M. The role of localized acidity generation in microbially influenced corrosion. *Curr. Microbiol.,* **2017**, *74*(7), 870-876.
[http://dx.doi.org/10.1007/s00284-017-1254-6] [PMID: 28444419]

[104] Olsson, C.O.A.; Landolt, D. Passive films on stainless steels—chemistry, structure and growth. *Electrochim. Acta,* **2003**, *48*, 1093-1104.
[http://dx.doi.org/10.1016/S0013-4686(02)00841-1]

[105] Bang, C.; Schmitz, R.A. Archaea associated with human surfaces: not to be underestimated. *FEMS Microbiol. Rev.,* **2015**, *39*(5), 631-648.
[http://dx.doi.org/10.1093/femsre/fuv010] [PMID: 25907112]

[106] Gupta, R.S. Protein phylogenies and signature sequences: A reappraisal of evolutionary relationships among archaebacteria, eubacteria, and eukaryotes. *Microbiol. Mol. Biol. Rev.,* **1998**, *62*(4), 1435-1491.
[http://dx.doi.org/10.1128/MMBR.62.4.1435-1491.1998] [PMID: 9841678]

[107] Li, X.; Liu, J.; Yao, F.; Wu, W.; Yang, S.; Mbadinga, S.M.; Gu, J.; Mu, B. Dominance of Desulfotignum in sulfate-reducing community in high sulfate production-water of high temperature and corrosive petroleum reservoirs. *Int. Biodeterior. Biodegradation,* **2016**, *114*, 45-56.
[http://dx.doi.org/10.1016/j.ibiod.2016.05.018]

[108] Völkl, P.; Huber, R.; Drobner, E.; Rachel, R.; Burggraf, S.; Trincone, A.; Stetter, K.O. Pyrobaculum aerophilum sp. nov., a novel nitrate-reducing hyperthermophilic archaeum. *Appl. Environ. Microbiol.,* **1993**, *59*(9), 2918-2926.
[http://dx.doi.org/10.1128/AEM.59.9.2918-2926.1993] [PMID: 7692819]

[109] Thauer, R.K.; Kaster, A-K.; Seedorf, H.; Buckel, W.; Hedderich, R. Methanogenic archaea: ecologically relevant differences in energy conservation. *Nat. Rev. Microbiol.,* **2008**, *6*(8), 579-591.
[http://dx.doi.org/10.1038/nrmicro1931] [PMID: 18587410]

[110] Stetter, K.O.; Lauerer, G.; Thomm, M.; Neuner, A. Isolation of extremely thermophilic sulfate reducers: evidence for a novel branch of archaebacteria. *Science,* **1987**, *236*(4803), 822-824.
[http://dx.doi.org/10.1126/science.236.4803.822] [PMID: 17777850]

[111] Beeder, J.; Nilsen, R.K.; Rosnes, J.T.; Torsvik, T.; Lien, T. Archaeoglobus fulgidus isolated from hot North Sea oil field waters. *Appl. Environ. Microbiol.,* **1994**, *60*(4), 1227-1231.
[http://dx.doi.org/10.1128/AEM.60.4.1227-1231.1994] [PMID: 16349231]

[112] Stetter, K.O.; Huber, R.; Blöchl, E.; Kurr, M.; Eden, R.D.; Fielder, M.; Cash, H.; Vance, I. Hyperthermophilicarchaea are thriving in deep North Sea and Alaskan oil reservoirs. *Nature,* **1993**, *365*, 743-745.
[http://dx.doi.org/10.1038/365743a0]

[113] Borenstein, S.W. *Microbiologically Influenced Corrosion Handbook*; Industrial Press Inc.: New York, **1994**.
[http://dx.doi.org/10.1533/9781845698621]

[114] Carew, J.; Al-Hashem, A.; El-Mohemeed, E.; Al-Enezi, H. North Kuwait oil field sea water flood experience in pipeline integrity management program. NACE Conference Papers (NACE International): TX, **2009**; pp. 1-8.

[115]   Gieg, L.M.; Jack, T.R.; Foght, J.M. Biological souring and mitigation in oil reservoirs. *Appl. Microbiol. Biotechnol.,* **2011**, *92*(2), 263-282.
[http://dx.doi.org/10.1007/s00253-011-3542-6] [PMID: 21858492]

[116]   Wu, Q.L.; Guo, W.Q.; Bao, X.; Yin, R.L.; Feng, X.C.; Zheng, H.S.; Luo, H.C.; Ren, N.Q. Enhancing sludge biodegradability and volatile fatty acid production by tetrakis hydroxymethyl phosphonium sulfate pretreatment. *Bioresour. Technol.,* **2017**, *239*, 518-522.
[http://dx.doi.org/10.1016/j.biortech.2017.05.016] [PMID: 28571628]

[117]   Sahrani, F.K.; Nawawi, M.F.; Usup, G.; Ahmad, A. Open circuit potential and electrochemical impedance spectroscopy studies on stainless steel corrosion by marine sulfate-reducing bacteria. *Malays. Appl. Biol.,* **2014**, *43*, 141-150.

[118]   Mah, T.F.C.; O'Toole, G.A. Mechanisms of biofilm resistance to antimicrobial agents. *Trends Microbiol.,* **2001**, *9*(1), 34-39.
[http://dx.doi.org/10.1016/S0966-842X(00)01913-2] [PMID: 11166241]

[119]   Xu, D.; Jia, R.; Li, Y.; Gu, T. Advances in the treatment of problematic industrial biofilms. *World J. Microbiol. Biotechnol.,* **2017**, *33*(5), 97.
[http://dx.doi.org/10.1007/s11274-016-2203-4] [PMID: 28409363]

[120]   Raad, I.; Chatzinikolaou, I.; Chaiban, G.; Hanna, H.; Hachem, R.; Dvorak, T.; Cook, G.; Costerton, W. *In vitro* and *ex vivo* activities of minocycline and EDTA against microorganisms embedded in biofilm on catheter surfaces. *Antimicrob. Agents Chemother.,* **2003**, *47*(11), 3580-3585.
[http://dx.doi.org/10.1128/AAC.47.11.3580-3585.2003] [PMID: 14576121]

[121]   Ballantyne, B.; Jordan, S.L. Biocides. In: *Pesticide toxicology and international regulation*; Marrs, T.C.; Ballantyne, B., Eds.; Wiley: Chichester, **2004**; pp. 384-385.

[122]   Wen, J.; Zhao, K.; Gu, T.; Raad, I.I. Chelators enhanced biocide inhibition of planktonic sulfate-reducing bacterial growth. *World J. Microbiol. Biotechnol.,* **2010**, *26*, 1053-1057.
[http://dx.doi.org/10.1007/s11274-009-0269-y]

[123]   Wen, J.; Zhao, K.; Gu, T.; Raad, I.I. A green biocide enhancer for the treatment of sulfate-reducing bacteria (SRB) biofilms on carbon steel surfaces using glutaraldehyde. *Int. Biodeterior. Biodegradation,* **2009**, *63*, 1102-1106.
[http://dx.doi.org/10.1016/j.ibiod.2009.09.007]

[124]   Yang, D. Mechanism and mitigation of biocorrosion by nitrate reducing pseudomonas aeruginosa against stainless steel. *Electronic Thesis or Dissertation.,* **2016**, https://etd.ohiolink.edu/

[125]   Kolodkin-Gal, I.; Romero, D.; Cao, S.; Clardy, J.; Kolter, R.; Losick, R. D-amino acids trigger biofilm disassembly. *Science,* **2010**, *328*(5978), 627-629.
[http://dx.doi.org/10.1126/science.1188628] [PMID: 20431016]

[126]   Kao, W.T.K.; Frye, M.; Gagnon, P.; Vogel, J.P.; Chole, R. D-amino acids do not inhibit *Pseudomonas aeruginosa* biofilm formation. *Laryngoscope Investig. Otolaryngol.,* **2017**, *2*(1), 4-9.
[http://dx.doi.org/10.1002/lio2.34] [PMID: 28286870]

[127]   Cava, F.; Lam, H.; de Pedro, M.A.; Waldor, M.K. Emerging knowledge of regulatory roles of D-amino acids in bacteria. *Cell. Mol. Life Sci.,* **2011**, *68*(5), 817-831.
[http://dx.doi.org/10.1007/s00018-010-0571-8] [PMID: 21161322]

[128]   Lam, H.; Oh, D.C.; Cava, F.; Takacs, C.N.; Clardy, J.; de Pedro, M.A.; Waldor, M.K. D-amino acids govern stationary phase cell wall remodeling in bacteria. *Science,* **2009**, *325*(5947), 1552-1555.
[http://dx.doi.org/10.1126/science.1178123] [PMID: 19762646]

[129]   Xu, H.; Liu, Y. D-amino acid mitigated membrane biofouling and promoted biofilm detachment. *J. Membr. Sci.,* **2011**, *376*, 266-274.
[http://dx.doi.org/10.1016/j.memsci.2011.04.030]

[130]   Royet, J.; Dziarski, R. Peptidoglycan recognition proteins: pleiotropic sensors and effectors of

antimicrobial defences. *Nat. Rev. Microbiol.,* **2007**, *5*(4), 264-277.
[http://dx.doi.org/10 1038/nrmicro1620] [PMID: 17363965]

[131]   Yu, C.; Wu, J.; Contreras, A.E.; Li, Q. Control of nanofiltration membrane biofouling by Pseudomonas aeruginosa using d-tyrosine. *J. Membr. Sci.,* **2012**, *423*, 487-494.
[http://dx.doi.org/10 1016/j.memsci.2012.08.051]

CHAPTER 10

Power Plant Corrosion

S. Ramesh[1,*], N.V. Krishna Prasad[1], N. Suresh Kumar[2], K. Chandra Babu Naidu[1,*], M.S.S.R.K.N. Sarma[1], K. Venkata Ratnam[3], H. Manjunatha[3], B. Parvatheeswara Rao[4] and T. Anil Babu[1]

[1] *Department of Physics, GITAM Deemed to be University, Bangalore - 562163, Karnataka, India*

[2] *Department of Physics, JNTU College of Engineering Anantapur, Anantapuramu-515002, A.P., India*

[3] *Department of Chemistry, GITAM Deemed to be University, Bangalore - 562163, Karnataka, India*

[4] *Department of Physics, Andhra University, Visakhapatnam- 530003, India*

Abstract: Corrosion is recognized as a serious problem in power plants that generate electricity. Many power plants generating a huge amount of electricity are needed to be taken care of it. Otherwise, these will cause a serious damage to human life. Corrosion gives rise to wastage of material in huge quantities, failure of tubes, leakage of tubes, sudden shutdowns as well as a reduction in the lifetime of components. Also, it reduces the thermal and electrical efficiency of a power plant to a maximum extent leading to minimum maintenance, outage, and replacement of cost. In understanding this problem, the present chapter illustrates the corrosions that take place in the power plants and preventive measures to be taken to avoid huge destruction to the life on the earth and to the environment. This, in turn, reduces the maintenance cost and damage to the human life.

Keywords: Corrosion, Environment, Erosion, Hot Corrosion, Thermal Spraying.

1. INTRODUCTION

In general, it is a well-known fact to the majority of people that corrosion is nothing but rust. "Rust" is a name which is likely to be reserved for iron, whereas corrosion refers to a natural process of converting a refined metal into a more chemically stable form such as hydroxide, oxide, or sulfide. It is a process of continuous eradication of metals due to chemical reaction and/or electrochemical reaction with their surrounding environment. This can also take place in materials

---

[*] **Corresponding authors S. Ramesh and K. Chandra Babu Naidu:** Department of Physics, GITAM Deemed to be University, Bangalore-562163, Karnataka, India; Tel: +91- 90000 00664; E-mails: sramesh664@gmail.com and chandrababu954@gmail.com

N. Suresh Kumar, P. Banerjee, H. Manjunatha and K. Chandra Babu Naidu (Eds.)

that are not metals, which include ceramics or polymers; the term "degradation" is more common. It is a known fact that corrosion degrades the useful properties of materials like strength, appearance and permeability to liquids and gases. Many alloys and metals show a tendency of merging with oxygen and water contained in their surrounding environment and come back to their stable state. Steel and iron generally interact with their environment to return to their stable oxide states. Hence, corrosion engineering can be treated as the field devoted to the control as well as prevention of corrosion. Like natural disasters like earthquakes or changes in weather, corrosion leads to expensive and dangerous damage to automobiles, drinking water systems, home appliances, gas lines, buildings, bridges, and power plants [1]. Throughout the world, the major production of electricity has happened from thermal power plants, where the water is boiled with the help of coal and the generated steam. This steam gets condensed and returns to the boiler by passing through the exit end of the turbine having lower pressure [2].

The development of more advanced electrical devices requires electrical power for the running of a device. It is made energy source compulsory in contemporary industrial societies. Statistical estimation shows that almost 70 percent of electricity production is from fossil power plants, while 15 percent production is from nuclear power plants and 12 percent is from power plants of the hydraulic type, and the remaining production is from other energy sources in developed countries like the United States of America. So, most countries depend on fossil fuel power plants for energy generation [3].

The fossil fuels involve coal, oil, and natural gas. Coal is a natural fossil fuel which can be extracted easily from the earth's crust compared to oil and gas. Coal is a combustible black or brownish-black sedimentary rock formed as rock strata called coal seams. Heat is generated by burning fossil fuel while one-fourth of the primary energy of the world and two-fifth of the world's electricity is produced by coal supplies. Many of the industries which can manufacture iron and steel and also power plants burn coal. Coal is reported to be a contaminated fuel that contains varying amounts of sulphur , hydrogen , oxygen and nitrogen. It is also called ash and is said to have a complex nature [4]. The amount of coal used in Indian power stations has almost a very high amount of ash (50 percent) that contains hard quartz belonging to abrasive mineral species (15 per cent), which increases the coals erosion propensity [5]. Literature indicates that deposited materials on the fireside that belong to gas turbine surfaces and corrosion leading to many crucial problems. Power plant corrosion leads to continuous and prolonged maintenance, enhancement in operation cost, reduction in efficiency and constitutes risks in terms of safety to workers. Prevention of corrosion is very crucial in order to enhance power generation equipment. This means that the coating for protection should be used for establishing a successful installation of insulation.

2. TYPES OF POWER PLANT CORROSION

1.1. Oxide Corrosion

This is also called dry corrosion referring to an electrochemical process that takes place on metal surface due to an oxygen molecule dissolving in water. In other words, a chemical change in which valence electrons are lost from the atom. If lagging is not installed properly, or if a failure of the protective surface coating takes place or if it is not at all applied, this type of corrosion occurs. Fig. (**1**) shows the corrosion in iron [6].

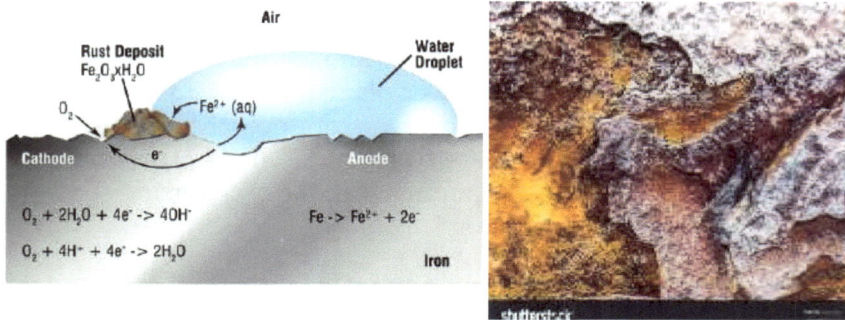

Fig. (1). Rust corrosion of Iron metal oxide [6].

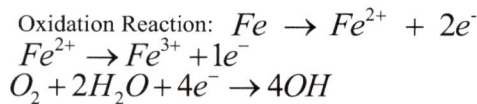

$$\text{Oxidation Reaction: } Fe \rightarrow Fe^{2+} + 2e^-$$
$$Fe^{2+} \rightarrow Fe^{3+} + 1e^-$$
$$O_2 + 2H_2O + 4e^- \rightarrow 4OH$$

The ferric ions combine with oxygen to form ferric oxide and are hydrated with variable amounts of water. In terms of Layman.

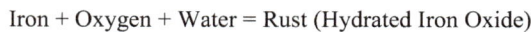

$$\text{Iron + Oxygen + Water = Rust (Hydrated Iron Oxide)}$$

2.2. Galvanic Corrosion

This corrosion occurs due to contact of two dissimilar metals giving rise to an electrical reaction which in turn leads to corrosion. The difference in electrical potential between the two metals produces a small current. The metal with lower electric potential more rapidly oxidizes in water. The conditions required for the galvanic corrosion are i) two dissimilar metals in contact, ii) oxygen concentration in water.

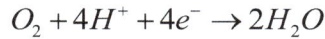

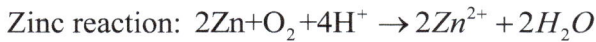

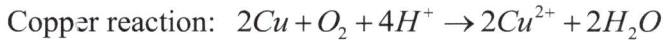

$$Zn \rightarrow Zn^{2+} + 2e^-$$
$$Cu \rightarrow Cu^{2+} + 2e^-$$
$$O_2 + 4H^+ + 4e^- \rightarrow 2H_2O$$

Zinc reaction: $2Zn + O_2 + 4H^+ \rightarrow 2Zn^{2+} + 2H_2O$

Copper reaction: $2Cu + O_2 + 4H^+ \rightarrow 2Cu^{2+} + 2H_2O$

The electric potential of copper is 1.90, whereas for zinc it is 1.65. Therefore, the flow of electrons will be from the zinc to the copper, which constrains the corrosion of copper. We know that the zinc corrodes at a faster rate than the copper surface, which is shown in Fig. (**2**).

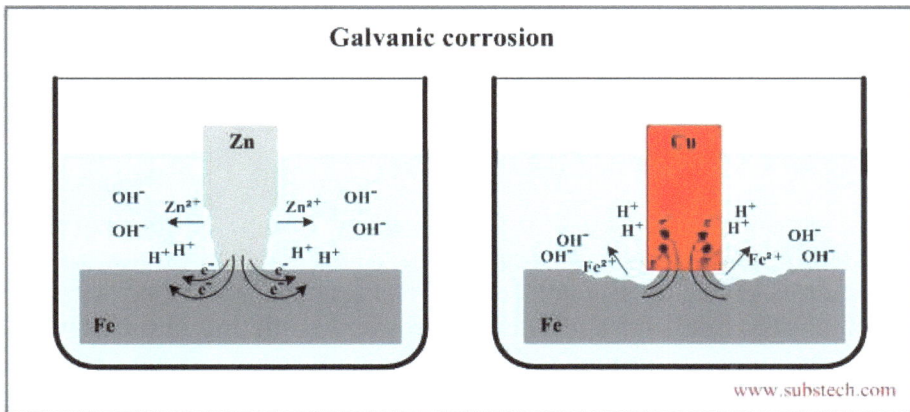

Fig. (2). Galvanic Corrosion in Zn and Cu [7].

Factors which determine galvanic corrosion are:

i) The electrode potential difference of two metals (higher the difference, higher will be the driving electron force of corrosion).

ii) The contact resistance that exists at the boundary between two metals (high contact resistance decreases corrosion rate and electric resistance of electrolyte solution by limiting the electron transfer through the boundary and dilute solutions with high electric resistance. It provides a low rate of corrosion.

iii) Ratio of the anode to cathode areas (Connecting large to small cathode results in a low rate of corrosion).

iv) Passive film presence.

v) Properties of electrolyte solution (Oxygen content, pH, temperature, and flow rate).

2.3. Hot Corrosion

Hot corrosion is formed due to a mechanism called accelerated corrosion. In addition, it is due to the existence of salt pollutants such as NaCl, V_2O_5 and Na_2SO_4 that form deposits of molten nature after combining. Further, it leads to damage of protective surface oxides. This can take place in diesel engines, gas turbines, furnaces or any other machinery having contact with hot gas consisting of some contaminants. The hot corrosion is generally formed due to two types of attacks given by (i) hot corrosion with high temperature (Type-I), and (ii) hot corrosion with low low-temperature (Type-II). Various parameters are likely to affect these two corrosions *viz.*, temperature, contaminant composition, alloy composition and flux rate, gas composition, velocity, and erosion process.

2.4. Type I Hot Corrosion (HTHC)

This can be observed mainly within the temperature range of 850-950°C [8]. Hot corrosion begins with alkali metal salts (fused) condensation on a component's surface. A series of chemically active reactions take place, starting with an attack of the protective oxide film and later progress by depleting chromium from the substrate. Depleting chromium leads to acceleration of oxidation in the base material and the formation of porous scale starts. Due to high thermodynamic stability for the hot corrosion, the Na_2SO_4 is considered as the effective salt. Sea salt contains NaCl and Na_2SO_4, which is the significant source of the sodium in the marine atmosphere. During the combustion reactions, sodium sulphate can form sodium and sulphate. Later it shows its presence in fuel. Vanadium, lead, chlorides, and phosphorus are the other impurities that are present in fuel or air which may form a salt mixture with lower melting temperature on combination with sodium sulphate which advances the attacking range [9, 10].

$$2NaCl + SO_2 + O_2 \rightarrow Na_2SO_4 + Cl_2$$

The potassium sulphate almost behaves similarly to the sodium sulphate with regards to hot corrosion. Thus, the alkali definition in air or fuel adds to form contents of sodium and potassium [11]. One of the contaminants that are unavoidable is vanadium in some liquid fuels. Deposits mixed with vanadium on exposure to high temperatures lead to hot corrosion.

2.5. Type II Hot Corrosion (LTHC)

LTHC corrosion is detected in the temperature range 650°C-800°C (which is < the melting point of Na_2SO_4 in the presence of a small amount of SO_3) [12]. Low melting point eutectic mixtures of Na_2SO_4 and $CoSO_4$ cause regular pitting in the localized areas in this type of corrosion. We normally notice the microscopic sulphidation and chromium depletion in low-temperature hot corrosion. This form of corrosion mainly occurs in gas turbines and industrial marine [13].

2.6. Mechanism of Hot Corrosion

Hot corrosion was identified to be one of the severe problems in the year 1940 due to the connection of its contact with fire-side boiler tubes. Its degradation in a plant generates steam, and it is coal fire based. Subsequently, this problem was noticed in gas turbines, boilers, internal combustion engines, fluidized bed combustion and industrial waste incinerators. It was identified to be very important in years the late 60s, with a gas turbine engine belonging to military aircraft suffering typical corrosion while operating over seawater during Vietnam War [14]. However, after different investigations, it was found that sulphide formation is a resultant of the reaction between the metal substrate and fused salt of sodium sulphate based thin film. This is named hot corrosion. Many mechanisms are developed to analyze the method of hot corrosion. Superalloys degradation generally consists of two phases. The first stage is called the initiation stage in which the alloys exhibit different behavior in the absence of the deposits. Besides, the second stage is called the breeding stage in which protective properties of oxide scales are caused by some deposits. These deposits become naturally more distinct than those with no presence of deposit [15]. Khana *et al.* [16], in their review, stated that alloys depend on selective oxidation which can resist corrosion to form thick Cr_2O_3 and Al_2O_3 scales of compact protective type. The biodegradation of the same materials takes place under hot corrosion conditions [16].

3. EROSION

It is also a type of corrosion occurred when high fluid surface velocities are exposed to aggressive chemical environments. Erosion is a very serious problem within turbines, pipelines, valves, heat exchanger tubes and combustion systems. In the case of coal fired boiler, fly ash particles can hit the boiler tubes and wear down. When the coal falls along the surface of the tubes and the corrosion takes initially in a steam generator tubes. Based on interacting substance and material surface, erosion wear may be divided into four different types, and they are solid particle erosion, slurry erosion, cavitation erosion and liquid impingement erosion.

3.1. Solid Particle Erosion (SPE)

This erosion process occurs due to loss of material by the regular strike of small, solid particles catch in air/gas at any important velocity. It is the main issue for the electric power industry, jet turbine, pipeline, and steam turbines. The estimated cost of these repairs is about US150 million a year. In 1981 macroscopic erosion mechanism was proposed by Bellman and Levy in which hitting of particles with material surface make platelet-like pieces and shallow craters. During the successive particle impacts, it is simple to separate these platelets from the material surface [17]. It is observed that during the development of platelets, adiabatic shear heating on the material surface and work hardenings will occur under the surface. On the other side, in the case of brittle materials, solid particle erosion is caused due to crack formation [18 - 24]. Radial and lateral cracks are formed when the solid particle impinges on the brittle surface. These developed cracks split the material surface into tiny parts, as shown in Fig. (3).

3.2. Cavitations Erosion (CE)

In 2006, Dular *et al.* [25] suggested the mechanism of cavitation erosion. This is a phenomenon in which fast changes of pressure in a liquid lead to the formation of small vapor-filled cavities in places where the pressure is relatively low. Cavitation erosion has been identified as one of the major problems in the design and operation of modern high-speed flow systems such as turbomachinery, fast nuclear reactors, flow control valves, high-speed mixing system and high-frequency ultrasonic devices.

3.3. Liquid Impingement Erosion (LIE)

It is a test for estimating the erosion resistance of certain materials. LIE is the progressive loss of original material forming a solid surface due to continued exposure to impacts by liquid drops or jets. It mainly occurs in elbows, 't' junction and the pipe wall downstream of the orifice in the pipelines. Herein, the flow velocity through the pipeline is highly increased causing the pipe wall thinning in power plants. It mainly depends on the flow velocity of the liquid [26].

3.4. Slurry Erosion (SE)

SE is a surface degradation process which arises due to repeated impacts of solid. This degradation due to SE depends on three main groups of factors associated with (i) Conditions of fluid flow which include flow velocity, the concentration of particles, particle impingement angle, density of the liquid, chemical activity of liquid and temperature of the liquid. (ii) Solid particles based on size, hardiness, shape, and strength. (iii) Target materials who's mechanical and endurance properties include toughness, yield, fatigue, strength, surface topography, work hardening, microstructure, number of defects and size of defects. Hence, the number of factors that influence slurry erosion is very high, and the degradation of materials is a synergic effect [27 - 31]. Fig. (4) shows the schematic representation of slurry corrosion and variation of erosion rate with impact angle.

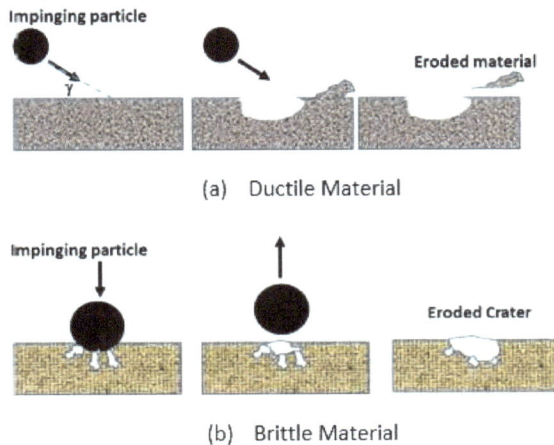

(a) Ductile Material

(b) Brittle Material

Fig. (3). (a) Erosion Procedure in Ductile Material & **(b)** Expected Mechanism of Erosion in Brittle Material (https://www.ansys.com/blog/erosion-fluid-dynamics-modeling).

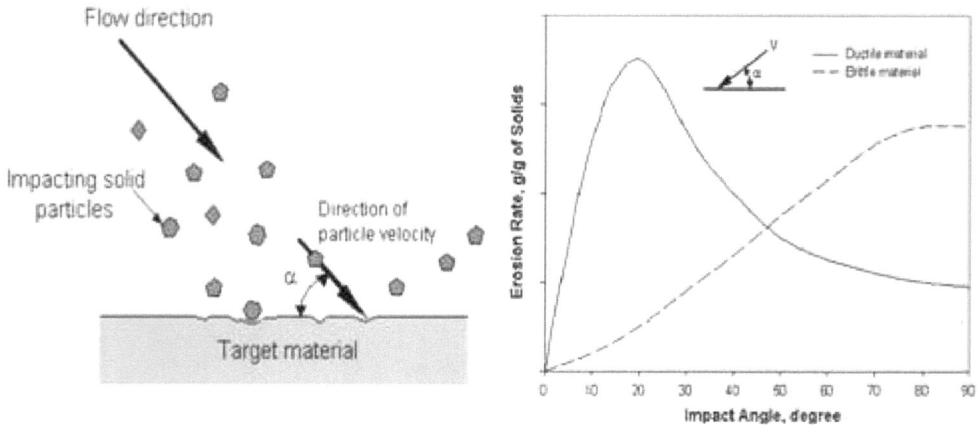

Fig. (4). Schematic diagram and erosion rate dependence on impact angle in slurry erosion [32].

The kinetic impact energy of single particles depends on the impact velocity and the size of the solid particle. With the increase in impact velocity, the size of erosion rate is increased. A similar effect on the rate of erosion increases the particle concentration in a liquid that has an influence on impact energy in total. Slurry Erosion is divided into low and high-velocity erosion depending on impact velocity. Increase in velocity above 6 to 9 m/s leads to the formation of high-velocity erosion, while the formation of low-velocity erosion takes place below this velocity [27].

4. PREVENTIVE METHODS OF CORROSION IN POWER PLANTS

Power plant corrosion gives rise to expensive repairs, maintenance of big length, losses of material, poor performance and if left unattended leads to failure. Industrial experts approved the corrosion prevention in the form of preventive and control strategies, such as regular inspections and the use of protective coatings. Power plant corrosion danger zones which need to be noticed to take preventive measures are like piping of hot and cold systems, boilers reactors, nacelles, turbines towers, plates, welding seams, boiler tubes, flue inlet gas ducts, scrubber outlet ducts, bypass ducts, modules, areas containing demineralized water, stacks, fuel handling areas, stack liners and collection sumps. Two practical solutions that control corrosion due to corrosive species which will enhance lifetime further regarding boiler components are either usage of highly alloyed materials or alleviating environment. These two methods are widely accepted.

4.1. Corrosion Resistant Materials and Alloys

The development of ultra-supercritical (USC) and AUSC advanced ultra-super critical (AUSC) dependent power plants fired by coal require very good performance steels and nickel-based superalloys [33]. The decrease in emission of CO_2 from coal-fired power plants can be attained with an increase in operating temperature of the steam system. This increases the overall efficiency of the power plant. In general, it is reported that a reduction of 3% CO_2 emission leads to 1% increase in absolute efficiency. Similarly, a decrease in 50% of CO_2 emission leads to an increase in efficiency from 36 to 55%. The more efficiency indicates that the more temperature and usage of medium like Cr-Ni, ferritic and ferritic-martensitic steels. They lead to higher corrosion rates in the steam atmosphere [34].

Steels like ferritic, austenitic grades and ferritic-martensitic test were carried in a closed loop and system of steam oxidation for temperatures in the range of 600 to 880°C. Studies of T. Dudziak, confirmed that T22 and T23 may come under the category of ferritic steels from thick oxide scales indicating the formation of three iron-based oxides at a temperature of more than 570°C. Formation of voids between Fe_2O_3 and Fe_3O_4 layers in ferritic materials shows oxide scale beyond 650°C, above which adding W of nearly 3 wt % to T23 steel enhances resistance and corrosion in a pure steam atmosphere as compared to T22 steel. Better corrosion resistance was shown when steel having 9 wt % Cr T91 and T92 in place of T22, T23 was used. Similarly, steels with more amount of Cr (16 wt % Cr, 316L steel) and steel with 17 wt % Cr and 18 wt % Cr show better corrosion resistance than that was offered by 12 wt % and 16 wt % Cr steel. Likewise, on comparison with ferritic steel, measurements of austenitic steel metal loss demonstrated that steel having 12 wt % chromium shows more than 5 μm metal loss while other materials show metal loss less than 5 μm after a huge exposure of 2000 hours. Very high alloyed steels like HR3C (25 wt % Chromium) 310S, 309S (greater than 22 wt % Chromium), exhibit creation of Cr_2O_3, Cr_3O_4, $MnCr_2O_4$ phases, protective scales, and reduction of spallation [33].

Corrosion related to materials used in construction metals is the main concern for operations in the geothermal industry (power plants). Technical challenges like scaling and its prevention need to be addressed by the industry. Selecting good material and efficient corrosion engineering increases the availability of power plants with a reduction in operating cost of the geothermal facilities. Taking economic aspects into consideration, components in contact with the processed brine replacement can be taken for approval. The approved processes are the usage of materials that are corrosion resistant throughout the industry or usage of materials with low cost. They may be replaced during failures. At the same time, materials of high quality are required for some of the process components, where the regular materials of conventional type fail due to lack of performance. This type of equipment is required to have high reliability and almost zero corrosion for performance, maintenance, and safety [35]. Tan *et al.* discussed the corrosion behavior of three Inconel alloys 617, 625, and 718. It was classified by displaying them to supercritical water (SCW) at 500°C and 25 MPa with ~ 25 ppb dissolved oxygen. GBE treatment was also performed on the alloy 617 to investigate its effect on corrosion behavior. By adjusting the grain boundary engineering (GBE) treatment, we can verify the optimized grain boundary character by electron backscattering diffraction. The GBE treated alloy of 617 samples exhibited the best corrosion resistance [36]. Usage of very cheap aluminide coating, which is easy to apply acts as one of the promising corrosion control techniques for superheater materials. On testing, it is found that corrosion resistance was of the order: Aluminide-coated Ni-based alloy with 59 > Aluminide-coated stainless steel with 310 > Ni –based alloy with 625 > Ni-based alloy with 59 > Fe based alloy with 556 > Stainless steel with 316. The hot corrosion resistance of FeAl and Fe_3Al formed on the low-cost stainless steel 310 is comparable to that of more expensive nickel-based alloys. Although the chlorination resistance is not particularly high, it is happened due to relatively low nickel content in the steel. The results of corrosion rates (in mm/year) against the position on the alloy are shown in Fig. (**5**) [37].

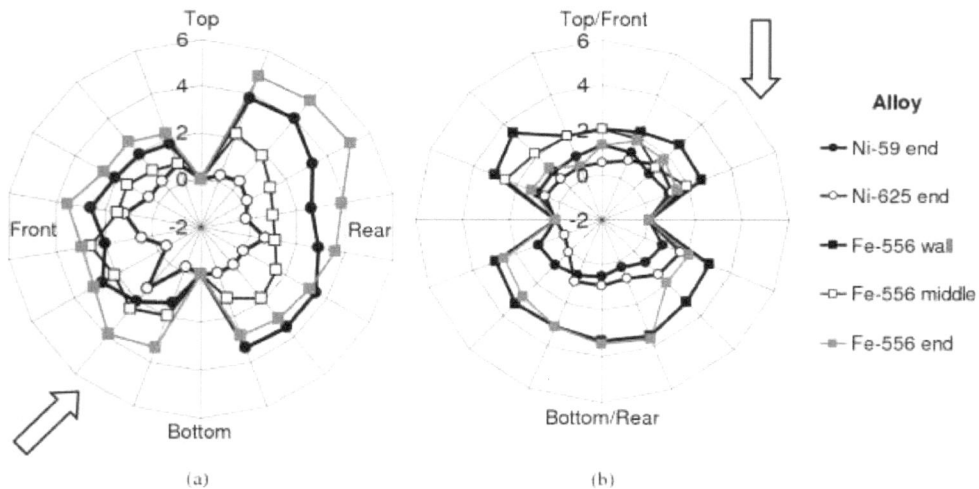

Fig. (5). Corrosion rates (mm/year) of alloy rings for sampling probe 1 and sampling probe 2 [37].

Autoclave exposure experiments were conducted with different alloyed materials and non-ferrous Titanium grade 2. Generally, the stainless steels and alloys show a very high resistance against uniform corrosion in the test environments. Most of the post exposure coupons did not show any macroscopic corrosion, and the surfaces were smooth and scaling free, as revealed in SEM Images. In case of the occurrence of no localized corrosion, uniform corrosion rates were below 0.005 mm/year. The past researchers mentioned that Cr, Ni, and Mo-alloys passivate during exposure in the geothermal brine. It is also mentioned that longer exposure increases the better results [38].

In nuclear power plants, copper is commonly used for the earthing system, due to its very high electrical conductivity and generally low corrosion rates in soil. Corrosion issues for copper earthing networks are corrosion of the copper material in soil along with effects of inhomogeneities, corrosion of earth electrodes from lighting strikes, and fault currents. The stray current corrosion of copper wires is induced by catholically protected objects or other sources, deteriorated electromechanical connections, corrosion of welds [39]. Kazutoshi *et al.*, developed a model of chemical effect on fast accelerated corrosion in power plants in consideration of the diffusion, if iron, chromium, dissolved hydrogen, and dissolved oxygen are under steady-state condition. The FAC rate shows a peak around 413 K, and it is decreased with an increase in Cr content of the material. It is also decreased with a rise in pH, and dissolved oxygen concentration. The thickness of the diffusion layer is considerably affected by fluid dynamics [40].

In nuclear power plants, alloy of steel ASTM, A-335Gr.P22 (21/4% Cr. 1% Mo) in the secondary circuit was used in place of severely degraded carbon steel pipe fittings. Further, the efficacy was observed on a trial basis. The fast-accelerated corrosion for a 90° elbow predicted the reduction to 1/10th of the FAC rates for carbon steel pipeline, as chromium is increased from 0.03 to 0.5%. The best grade of carbon steel (ASTM A.335 Grade 11/22) is used in many plants and hence, it provides better resistance to FAC. The surface oxide changes from Fe_3O_4 to $FeCr_2O_4$ with the inclusion of Cr in carbon steel. From the above, another replacement with the same grade A 106, grade B (containing 0.25-0.40% chromium) would be competent [41].

4.2. Modification of the Environment/Coatings

Coating technology is a technology which was enormously growing in the material science field. A layer of material is defined as coating naturally developed, synthetically developed, or artificially deposited on objects surface made with another material to achieve required technical properties [42, 43]. A combination of the development of materials is created for erosion and corrosion resistance. The appropriate approach for the application of these corrosion materials, as a coating would be the optimum solution. Almost all the recent studies exhibited that 80% of the total expenditures to safeguard metals are related to coating applications. Classification of coating systems is diffusion type or overlay type that distinguishes deposition method or structure of resulting coating substrate bond. From the manufacturing point of view, there are three methods which are currently used for the deposit coatings. Those are chemical vapor deposition (CVD), physical vapor deposition (PVD) and plasma spraying. The CVD comes under the diffusion coating, and PVD and thermal spraying process come under the class of overlay coatings, in which the desired material is placed over the substrate material [44]. Scientists showed that TiO_2 made of Al_2O_3 containing 13 wt % TiO_2 was used to enhance the wear corrosion and erosion of resistance of steel. Powder particles are injected into a plasma jet for plasma spray processing Al_2O_3 -13 wt % TiO_2 coatings thereby causing them to melt into droplets that are propelled towards the substrate. Solidification of these droplets build-up a coating, generally of 100-300 μm thick. Plasma sprayed zirconia coatings as thermal barrier coatings are enforced to hot section parts of gas engines to raise the temperature capability Ni-base superalloys [45, 46].

Researchers explored plasma-sprayed metallic coating of nickel-aluminide deposited on Fe-based super alloy. This coating exhibited better erosion resistance as compared to the uncoated samples. Scientists studied hot corrosion performance of plasma sprayed coatings on an iron-based superalloy. On F-based alloys, metallic coating of NiCrAlY, Ni_3Al, Ni-20Cr, and Stellite-6 were deposited. NiCrAlY is used as a bond coating in all cases [47, 48]. These overlay coatings exhibited better resistance than uncoated steel to hot corrosion. NiCr coating followed by Cr_2C_3-NiCr coating was found to be most protective. Least effective method to protect the substrate steel is WC-Co coating.

Kalss *et al.*, considered aluminium and titanium-based coating and explained that TiAl based nitrides which are like TiAlN & AlTiN are stable against oxidation up to a temperature of 800°C. AlCrN is the best oxidation resistance coating material. Even at 1100°C, only a thin layer of about 150 nanometers in thickness could be observed [49]. Tsipas & Pereza *et al.*, studied the aluminium manganese coatings which are deposited by CVD-FBR on P-92 and HCM12 steels as a protective layer. It is a powerful method to obtain Mn-containing aluminide coatings on ferrite steels. The heat-treated aluminium coated AISI 304 specimens may find an application due to the combination of their toughness and the potential corrosion properties [50, 51].

CONCLUSION

Corrosion is a continuous natural process; it will never stop, but can be minimized. Most of the ores mined into metals like sulphides, oxide, carbonates and are handled into metals twhich we use in our electrical systems. Almost round the clock, these metals try to come back to their most stable state. The ores are mined during the corrosion process. While the science of corrosion in power plants is quite known and understood, limiting the impact of corrosion is much of an art. The corrosion in power plants can be protected in understanding the underlying principles of corrosion along with apparatus to be preserved, the installation environment, material selection, coating selection and sheltering equipment.

CONSENT FOR PUBLICATION

Not applicable.

CONFLICT OF INTEREST

The author declares no conflict of interest, financial or otherwise.

ACKNOWLEDGEMENTS

Declared none.

REFERENCES

[1]   Khanna, A.S. *Introduction to high temperature oxidation and corrosion*; ASM International, **2002**, p. 324.

[2]   Stringer, J. High temperature corrosion problems in coal-based power plant and possible solutions **1997**.

[3]   Viswanathan, R. *Damage mechanism and life assessment of high-temperature components*; ASM International, **1989**, pp. 1-483.

[4]   Natesan, K. "Corrosion performance of materials in coal-fired power plants", Proceedings International conference on corrosion 'CONCORN' 97, December 3-6, Mumbai, India, 1997. 24-35.

[5]   Krishnamoorthy, P.R.; Seetharamu, S.; Sampathkumaran, P. Influence of the mass flux and impact angle of the abrasive on the erosion resistance of materials used in pulverized fuel bends and other components in thermal power stations. *Wear,* **1993**, *165*, 151-157.
        [http://dx.doi.org/10.1016/0043-1648(93)90330-O]

[6]   https://www.shutterstock.com/image-photo/rust-corrosion-iron-metal-oxide-texture-762675280

[7]   https://www.substech.com/dokuwiki/doku.php?id=galvanic_corrosion

[8]   Hancock, P. Vanadic and chloride attack of supperalloys. *Mater. Sci. Technol.,* **1987**, *3*(7), 536-544.
        [http://dx.doi.org/10.1080/02670836.1987.11782265]

[9]   Jaffee, R.I.; Stringer, J. High-temperature oxidation and corrosion of superalloys in the gas turbine (a review). In: *Source book on materials for elevated-temperature applications*; Bradley, F., Ed.; ASM: Metals Park, **1979**; pp. 19-33.

[10]   Sanorelli, R.; Sivieri, E.; Reggiani, R.C. High-temperature corrosion of several commercial Fe-Cr-Ni alloys under a molten sodium sulphate deposit in oxidizing gaseous environments. *Mater. Sci. Eng. A,* **1989**, *120*, 283-291.
        [http://dx.doi.org/10.1016/0921-5093(89)90752-1]

[11]   Wright, I.G. *High–temperature oxidation and corrosion*; , **1987**.

[12]   Stringer, J. *High Temperature Corrosion of Superalloys,* **1987**.
        [http://dx.doi.org/10.1080/02670836.1987.11782259]

[13]   Meier, H.G. A review of advances in high-temperature corrosion. *Mater. Sci. Eng.,* **1989**, *120*, 1-11.

[14]   Khana, A.S.; Jha, S.K. Degradation of materials under hot corrosion conditions. *Trans. Indian Inst. Met.,* **1998**, *51*(5), 279-290.

[15]   Rapp, R.A. Hot corrosion of materials: a fluxing mechanism. *Corros. Sci.,* **2002**, *44*, 209-221.
        [http://dx.doi.org/10.1016/S0010-938X(01)00057-9]

[16]   Vikas Chawla, S.; Prakash, D.P.; Buta, S.; Manoj, S. Hot corrosion-a review. *Global Conference on Production & Industrial Engineering,* **2007**.

[17]   Bellman, R.; Levy, A. Erosion mechanism in ductile metals. *Journal of Wear,* **1981**, *70*, 1-28.
        [http://dx.doi.org/10.1016/0043-1648(81)90268-4]

[18]   Chase, D.; Rybicki, E.; Shadley, J. A model for the effect of velocity on erosion of N80 steel tubing due to the normal impingement of solid particle. *J. Energy Resour. Technol.,* **1992**, *114*, 54-64.
        [http://dx.doi.org/10.1115/1.2905921]

[19]   Hutchings, I. Some comments on the theoretical treatment of erosive particle impacts. *Proc. of the 5th Int. Conf. on Erosion by Liquid and Solid Impact,* **1980**, pp. 36-41.

[20]   Andrews, D. An analysis of solid particle erosion mechanisms. *J. Phys. D Appl. Phys.,* **1981**, *14*, 1979-1991.
        [http://dx.doi.org/10.1088/0022-3727/14/11/006]

[21]   Jahanmir, S. The mechanics of subsurface damage in solid particle erosion. *Wear,* **1980**, *61*, 309-338.
        [http://dx.doi.org/10.1016/0043-1648(80)90294-X]

[22]   Srinivasan, S.; Scattergood, R.O. Effect of erodent hardness on erosion of brittle materials. *Wear,* **1988**, *128*(2), 139-152.
        [http://dx.doi.org/10.1016/0043-1648(88)90180-9]

[23]   Sundararajan, G. A comprehensive model for the solid particle erosion of ductile materials. *Wear,* **1991**, *149*, 111-127.
        [http://dx.doi.org/10.1016/0043-1648(91)90368-5]

[24]   Kleis, I.; Kulu, P. *Solid Particle Erosion Occurrence, Prediction and Control;* , **2008**.

[25]   Dular, M.; Stoffel, B.; Sirok, B. Development of a cavitation erosion model. *Wear,* **2006**, *261*, 642-655.
        [http://dx.doi.org/10.1016/j.wear.2006.01.020]

[26]   Fujisawa, N.; Wada, K.; Yamagata, T. Numerical analysis on the wall- thinning rate of bent pipe by liquid droplet impingement erosion. *Eng. Fail. Anal.,* **2016**, *62*, 306-315.
        [http://dx.doi.org/10.1016/j.engfailanal.2016.01.005]

[27]   Grewal, H.S.; Agrawal, A.; Singh, H. Design and development of high-velocity slurry erosion test rig using CFD. *J. Mater. Eng. Perform.,* **2013**, *22*, 152-161.
        [http://dx.doi.org/10.1007/s11665-012-0219-y]

[28]   Zbrowski, A.; Mizak, W. Analiza systemówwy korzystywanych w badan iachud erzeni oweg ozużyciae rozyj nego. *Proble my ekspl oata cji,* **2011**, *3*, 235-250.

[29]   Finnie, I. Some reflections on the past and future of erosion. Wear 186-187 (1995) 1-10. 10. Finnie I.: Erosion of surfaces by solid particles. *Wear,* **1960**, *3*, 87-103.
        [http://dx.doi.org/10.1016/0043-1648(60)90055-7]

[30]   Grewal, H.S.; Agrawal, A.; Singh, H. Slurry erosion mechanism of hydroturbine steel: Effect of operating parameters. *Tribol. Lett.,* **2013**, *52*, 287-303.
        [http://dx.doi.org/10.1007/s11249-013-0213-z]

[31]   Lathabai, S.; Pender, D.C. Microstructural influence in slurry erosion of ceramics. *Wear,* **1995**, *189*, 122-135.
[http://dx.doi.org/10.1016/0043-1648(95)06679-9]

[32]   Arora, M.; Ohl, C.D.; Morch, K.A. *Cavitation inception on microparticles: A self-propelled particle accelerator.,* **2014**.

[33]   Dudziak, T. *Steam Oxidation of Fe Based Materials,*
[http://dx.doi.org/10.5772/62935]

[34]   Henry, J.; Zhou, G.. Ward, T. Lessons from the past: materials-related issues in an ultra-supercritical boiler at Eddystone plant. *Mater. High Temp.,* **2007**, 24.
[http://dx.doi.org/10.3184/096034007X277924]

[35]   Huttenloch, P.; Sanjuan, B.; Kohl, T.; Steger, H.; Zorn, R. Corrosion and scaling as interrelated phenomena in an operating geothermal power plant. *Corros. Sci.,* **2013**, *70*, 17-28.
[http://dx.doi.org/10.1016/j.corsci.2013.01.003]

[36]   Tan, L.; Ren, X.; Sridharan, K.; Allen, T.R. Corrosion behaviour of Ni-base alloys for advanced high temperature water-cooled nuclear plants. *Corros. Sci.,* **2008**, *50*, 3056-3062.
[http://dx.doi.org/10.1016/j.corsci.2008.08.024]

[37]   Awassada, P.; Changkook, R.; Yao, B.Y.; Karen, N. Investigation into high temperature corrosion in a large municipal waste - to - energy plant. *Corros. Sci.,* **2010**, *52*, 3861-3874.
[http://dx.doi.org/10.1016/j.corsci.2010.07.032]

[38]   Mundhenk, N.; Huttenloch, P.; Sanjuan, B.; Kohl, T.; Steger, H.; Zorn, R. Corrosion and scaling as interrreleated phenomena in an operating geothermal power plant. *Corros. Sci.,* **2013**, *70*, 17-28.
[http://dx.doi.org/10.1016/j.corsci.2013.01.003]

[39]   Burda, P.A. Differential aeration effect on corrosion of copper concentric neutral wires in the soil. In: *Effects of Soil Characteristics on corrosion", ASTM STP 1013*; Chaker, V.; Palmer, J.D., Eds.;
[http://dx.doi.org/10.1520/STP19707S]

[40]   Fujiwara, K.; Masafumi, D.; Kimitoshi, Y.; Fumio, I. Model of physic-chemical effect on flow accelerated corrosion in power plant. *Corros. Sci.,* **2011**, *53*, 3526-3533.
[http://dx.doi.org/10.1016/j.corsci.2011.06.027]

[41]   Vivekanand, K. Flow accelerated corrosion: forms, mechanisms and case studies. *Procedia Eng.,* **2014**, *86*, 576-588.
[http://dx.doi.org/10.1016/j.proeng.2014.11.083]

[42]   Sidhu, B.S.; Prakash, S. Evaluation of the behavior of shrouded plasma spray coatings in the platen superheater of coal-fired boilers. *Metallurgical & Materials Transactions,* **2006**, *37*, 1927.
[http://dx.doi.org/10.1007/s11661-006-0135-6]

[43]   Sidhu, T.S.; Agarwal, R.D.; Prakash, S. Hot corrosion of some super alloys and role of high-velocity oxy-fuel spray coatings. *Surf. Coat. Tech.,* **2005**, *198*, 441-446.
[http://dx.doi.org/10.1016/j.surfcoat.2004.10.056]

[44]   Khanna, A.S. *Introduction to High Temperature Oxidation and Corrosion*; ASM International, **2002**, p. 34.

[45]   Bansal, P.; Padture, N.P.; Vasiliev, A. Improved interfacial properties of Al2O313wt%TiO2 plasma sprayed coatings derived from nanocrystalline powders. *Acta Mater.,* **2003**, *51*, 2959-2970.
[http://dx.doi.org/10.1016/S1359-6454(03)00109-5]

[46]   Thermal shock resistance of nano-structured and conventional zirconia coatings deposited by atmospheric spraying. *Surface & Coating Technology Journal,* **2005**, *197*, 85-192.

[47]   Mishra, S.B.; Chandra, K.; Prakash, S.; Venkataraman, B. Characterization and erosion behaviour of a plasma sprayed Ni3Al coating on a Fe-based superalloy. *Mater. Lett.,* **2005**, *59*, 3694-3698.
[http://dx.doi.org/10.1016/j.matlet.2005.06.050]

[48] Harpreet Singh, D. Some studies on hot corrosion performance of plasma sprayed coatings on a Fe-based superalloy", 2005. 192, 27-38.

[49] Kalss, W.; Reiter, A.; Derfinger, V.; Gey, C.; Endrino, J.L. Modern coatings in high performance cutting applications. *Int. J. Refract. Met. Hard Mater.,* **2006**, *24*, 399-404. [http://dx.doi.org/10.1016/j.ijrmhm.2005.11.005]

[50] Tsipas, S.; Brossard, J.M.; Hierro, M.P.; Trilleros, J.A.; Sánchez, L.; Bolívar, F.J.; Pérez, F.J. Al–Mn CVD-FBR protective coatings for hot corrosion application. *Surf. Coat. Tech.,* **2007**, *201*, 4489-4495. [http://dx.doi.org/10.1016/j.surfcoat.2006.06.023]

[51] P'ereza, F.J. U, F. Pedrazaa, M.P. Hierroa, P.Y. Hou, "Adhesion properties of aluminide coatings deposited *via* CVD in fluidised bed reactors_CVD-FBR/on AISI 304 stainless steel. *Surf. Coat. Tech.,* **2000**, *133-134*, 338-343. [http://dx.doi.org/10.1016/S0257-8972(00)00952-X]

CHAPTER 11

Corrosion in Chemical and Fertilizer Industries

N.V. Krishna Prasad[1,*], **S. Ramesh**[1], **K. Chandra Babu Naidu**[1], **M.S.S.R.K.N. Sarma**[1], **K. Venkata Ratnam**[2], **H. Manjunatha**[2] and **B. Chandra Sekhar**[3]

[1] *Department of Physics, GITAM Deemed to be University, Bangalore - 562163, Karnataka, India*

[2] *Department of Chemistry, GITAM Deemed to be University, Bangalore - 562163, Karnataka, India*

[3] *Vignan's Institute of Engineering for Women, Visakhapatnam- 530003, India*

Abstract: Any country's economy and its development primarily depend on its infrastructure apart from existing natural resources in that region. The infrastructure mainly refers to irrigation system, buildings, roads, bridges, airports, transport, education and industries located there. Here it is noteworthy that all these mentioned infrastructures will be corrosion affective which may undergo degradation and deterioration processes. Corrosion is an unavoidable problem that mainly impacts industrial environment. Anthropogenic activity worldwide leads to enhancement of atmospheric pollution which indirectly accelerates corrosion in the form of rust, water pollution. Major industries in any country relate to chemical and fertilizers. India is country with dense population and developing industrially at vast rate. Chemical industry includes companies producing industrial chemicals in which raw materials like water, air, oil, natural gas, minerals and metals converted into seventy thousand products of different type. Indian statistics for the year 2018-19 indicate a production of major petrochemicals and chemicals at 27,847 MT whereas 27,735 MT during 2017-18. In this chapter, we mainly focus on corrosion related to chemical and fertilizer industries, impact of corrosion on their efficiency, corrosion controlling methods and their interrelated phenomena if any.

Keywords: Chemicals, Corrosion, Fertilizers, Industry.

1. INTRODUCTION

Deterioration of a given material on reacting with its surrounding environment is known as corrosion which is a process of converting a refined metal naturally into a hydroxide, oxide or sulfide. It gradually destructs metals by undergoing electrochemical or chemical reaction with surrounding environment [1].

* **Corresponding author H. Manjunatha:** Department of Chemistry, GITAM Deemed to be University, Bangalore-562163, Karnataka, India; Tel: +91- 88611 58964; E-mail: hanumanjunath80@gmail.com

N. Suresh Kumar, P. Banerjee, H. Manjunatha and K. Chandra Babu Naidu (Eds.)

Corrosion is a continuous process and every metal wherever used undergoes some form of corrosion. Corrosion in gas and oil industry is due to water, hydrogen sulfide, carbon dioxide and other existing microbiological activities. If not controlled corrosion in industrial sector may lead to serious problems. Corrosion is a problem of danger and also expensive. Corrosion leads to collapse of bridges, buildings, leakage of oil pipelines and chemical plants, *etc*. Major corrosion results from reactions electrochemically while general corrosion occurs due to atomic oxidization on same metal surface damaging the entire surface. Generally majority of metals easily oxidize due to loss of electrons to oxygen and form an oxide with metal. Process of reduction and oxidation taking place on different types of metals in contact is known as galvanic corrosion. In electronic equipment water or moisture becomes trapped between two electrical contacts having electrical voltage between them. This process is known as electrolytic corrosion. This results in an unintended electrolytic cell. Harmful corrosion can be prevented in many ways. Some metals naturally resist corrosion by reacting with corrodes in the oxygen in air [2 - 5].

This results in formation of a thin oxide film which blocks the tendency of metal to undergo further reaction. Rust, Galvanic corrosion, Stress cracking corrosion, General corrosion, localized corrosion and Caustic agent corrosion are the six types of corrosions given below:

2. TYPES OF CORROSION

2.1. Rust

Rust is a simple example of corrosion which is considered to be a result of oxidation. It is of importance that not all iron oxides are rust. Rust may be formed when oxygen reacts with iron but at the same time keeping iron and oxygen together is not sufficient to form rust. Fig. (**1**) shows the rust in iron.

2.2. Galvanic Corrosion

Galvanic corrosion is also known as bimetallic corrosion. It is due to an electro-chemical process in which two metals are in electrical contact in the presence of an electrolyte where in one metal corrodes. The galvanic corrosion is shown in Fig. (**2**).

Fig. (1). Rust (Courtesy: www.shutterstock.com/).

Fig. (2). Galvanic corrosion (Courtesy: www.azom.com).

2.3. Stress Corrosion Cracking

Enhancement or growth of crack formation in a corrosive environment is known as stress corrosion cracking (SCC) as shown in Fig. (**3**). It may give rise to sudden failing of normal ductile metal alloys if subjected to tensile stress especially at higher temperatures.

Fig. (3). Stress corrosion cracking (Coutesy:https://www.met-tech.com/).

2.4. General Corrosion

General corrosion is a corrosion that forms at same rate on the entire surface of the metal when exposed to conditions that cause corrosion. These affected surfaces due to this process may have areas with more or less penetration. Fig. (**4**) illustrates the schematic representation of general corrosion.

Fig. (4). General corrosion (Courtesy:www. nace.org).

2.5. Localized Corrosion

Localized corrosion is a corrosion in which localized sites on the metal surface face an intense attack. An example for localized corrosion is shown in Fig. (**5**).

Fig. (5). Localized corrosion (Courtesy: https://www.nitty-gritty.it/localized-corrosion).

2.6. Caustic Corrosion

Caustic agents present in products related to household give rise to caustic corrosion shown in Fig. (**6**). Properties of corrosion and ingested agent concentration decide the severity of damage. Caustic agents responsible for this type of damage are drain cleaners and lye soaps which come under the category of strong alkaline products used for cleaning purpose [6 - 11].

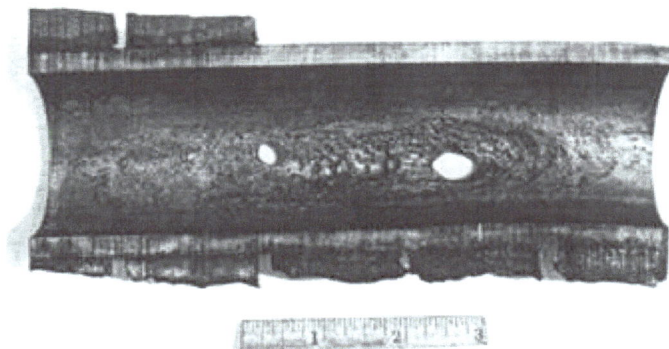

Fig. (6). Caustic corrosion (Courtesy: Chapter 3 of Boiler tube failures).

3. CHEMICAL FERTILIZER INDUSTRY AND ITS GROWTH IN INDIA

Chemical fertilizers play an important and significant role in the success of Indian Green Revolution. It also plays vital role in food and grain production. In view of this, Indian Government has been pursuing the policies related to enhance the consumption and availability of fertilizers in India on a consistent basis. India is said to be achieved almost self-sufficiency in producing urea and could manage our requirement of nitrogenous fertilizers through the natural industry [11]. Coming to phosphate fertilizers almost 50% of the domestic requirement is satisfied through natural production while the required intermediate materials and raw materials are majorly imported. Since potash has no viability in terms of source or reserves in India it is totally imported. Our Indian industry of fertilizers began in 1906 with first ever manufacturing unit comprising of Single Super Phosphate at Ranipet near to Madras (Chennai) having 6000 Metric Tonnes capacity annually. Subsequently in 1940's and 1950's FACT (Fertilizer and Chemicals Travancore of India) was started at Kochi of Kerala while FCI(Fertilizers Corporation of India) was set up in Sindri of Jharkhand (Bihar at that time) which were said to be the first large sized fertilizer plants and expected to achieve self-sufficiency in food-grains. Subsequent green revolution in late 60's spiked the fertilizer industry growth in India leading to huge production in fertilizers during 70's and 80's.

4. CORROSION IN FERTILIZER INDUSTRY

Corrosion is one of the major problems in fertilizer plants due to the presence of a wide spectrum of corrosives as raw materials like natural gas, naptha, fuel oil, coal, rock phosphate, sulphur, sulphuric acid, nitric acid, phosphoric acid *etc*. Especially phosphatic and nitrogenous fertilizer production units are said to face the effect of corrosion due to usage of raw materials in the process of production which has corrosive nature. Heavy risk has been encountered by the damage caused due to corrosion leading to heavy loss by this industry. Previous work reviewed different forms of corrosion observed in fertilizer plants and recommended certain measures in order to improve reliability of the overall plant. Indian agriculture is mainly dependent on phosphatic fertilizers which include single super phosphate, triple super phosphate and nitrogenous fertilizers that include ammonium nitrate, ammonium sulphate, calcium nitrite di-ammonium phosphate and urea [12].

5. CORROSION IN UREA MANUFACTURING PLANTS

Corrosion can be determined with the help of temperature, concentration of dissolved oxygen and process components [13]. Corrosion may be accelerated due to existence of contaminants. Usage of SS (Stainless Steel) lining made vessels of carbon steel in HP synthesis section and leak detection units (1, 2) show bad effect on process units which take help of chemicals. Specific types of corrosion that may take place in urea plants are:

1. Active corrosion which is a general form of corrosion. This may be reduced by exposing stainless steel to carbonate solution and make unreactive by altering the surface layer or coating the surface with a thin inert layer. If not, equipment longevity will be reduced because of stainless steel suffering from severe thinning.

2. Inter Granular Corrosion (IGC) which is corrosion because of oxidizing action. Action between urea carbonate solution with oxygen, low ammonia/carbon dioxide ratio and separation of impurities in sensitized stainless steel.

3. Erosion corrosion (A geological process where earthen materials are worn away and get transported due to by natural forces like water or wind).

4. Galvanic corrosion occurring in CO_2 gas cooler cleaning circuit (3) which is reported to result in complete plant shutdown over a period of six months.

5. Stress Corrosion Cracking (SCC) due to chloride action in stripper/carbonate condenser shell.

6. CORROSION IN PAPER INDUSTRY

Corrosion in paper industry is related to corrosion that occurs in the infrastructure used for pulping of paper. Technically paper industry progressed in terms of minimizing corrosion. Bronze, stainless steel and cast iron are the materials used in infrastructure of paper industry which witnessed high rate of failure due to corrosion. White water equipment subjected to aggressive environment in papermaking machinery corrosion occurs [14]. Exposure of metal surfaces to white water is required in this industry. It will steam the crack formation favoring the deposit of pulp and other compounds. Equipment made of stainless steel has good general corrosion resistance and weldability. Usage of oxygen and chlorine gas in bleach plant and pulp bleaching favors aggressive oxidant and stainless steel. Also, more corrosion is reported in these plants due to the washing systems used for paper pulp. Hence paper industry opts for usage of nickel, high alloy stainless steel and titanium for good resistance to corrosion in a particular environment [15]. Stainless steel when exposed to corrosive environment due to bleach plants get benefited by nickel, molybdenum and chromium being alloy elements and have resistance increase which initiate crevice and pitting corrosion. Addition of nitrogen will increase the resistance to pitting corrosion when molybdenum is added. Corrosion is also said to occur in the pulping facility of liquor by suction rolls, chemical recovery boilers and sulfites.

7. CORROSION IN OIL AND GAS INDUSTRY

Gas and oil industries commonly witness problems related to corrosion. Oil pipelines, gas pipelines and pipelines used in refinery as well as petrochemical plant encounter severe problems of corrosion caused by water, H_2S and CO_2 in oil, gas industries is known to be corrosion internal. Microbiological activity aggravates this corrosion [1]. It is important that multiphase fluids flow heavily influence corrosion rate. Corrosion which is flow induced and corrosion due to erosion might occur at high rate of flow whereas corrosion due to pitting occurs at low flow rate. Nature of sediments and amount will decide corrosion since velocity with high flow sweeps the sediments outside while flow with less velocity makes the sediments to settle at the bottom of the pipelines giving chance to produce space for pitting corrosion [2]. Economically speaking corrosion with respect to combating in gas and oil industry is significant in terms of loss due to corrosion is very high [3]. One method of solving this economic problem is to use corrosion inhibitor [4, 5]. These inhibitors can be categorized into cathodic, anodic inhibitors or mixed corrosion inhibitors. Chemical nature decides whether they are organic inhibitors or inorganic inhibitors. Inhibitor chemisorption is the general mechanism in which the inhibitors surface of the metal forms a thin film for protection which in turn protects the metal that is underlying from effect of corrosion. Corrosion inhibitors used in this industry which are mixtures containing emulsifiers, surfactants, enhancers come under commercial inhibitors [16]. To use these inhibitors in an effective manner they must be environment compatible, economically compatible up to expectation and should not cause any important unnecessary side effects that affect the operation process in addition to providing high protection to the metal. One important property of inhibitor is it should be thermally stable. In addition, mixing effect of inhibitor with surrounding environment like water tolerance, solubility, emulsion, formation of foam and viscosity, drying, density and pour point (physical properties) should be considered. Very often these inhibitors get diluted before injection and improve mobility under cold weathers. It is difficult to enhance the equipment life, prevention of accidents and shutdowns because of catastrophic mechanical failures in this industry. It is critical to avoid product contamination as well as heat transfer loss prevention [17, 18]. Many other costs are associated in using inhibitors apart from the installation and maintenance costs for equipment installation.

8. CORROSION DUE TO AGRICULTURAL CHEMICALS

Commercial chemicals used in farming which include fertilizers, preservatives, pest chemicals, grains and farm wastes are also corrosive. Fertilizers are chemicals that promote plant growth [19]. Generally, they are applied through soil or by spraying. Fertilizers are used to provide three major nutrients like potassium, nitrogen and phosphorus and secondary nutrients like sulphur, calcium, magnesium and iron in required proportions. Corrosion between fertilizers varies specially based on decomposition or producing aggressive substances through reactions given by hydrogen sulphide or ammonia. No corrosion occurs if the fertilizers are kept in dry place but due to hygroscopic nature, they collect moisture and become corrosive. Based on the relative humidity at which moisture is absorbed corrosion varies from one compound to another [20]. Ammonium nitrate absorb moisture at 60% RH where as some of the phosphates absorb moisture only above 90% RH. Moisture initially causes caking of the fertilizer, which can increase its abrasive properties [5].

CONCLUSION

In view of the challenges being faced by chemical and fertilizer industry it is strongly recommended that a methodology should be adopted by fertilizer industry to have good maintenance and inspection optimization programs for meeting the risk of high corrosion that affects the plant equipment. Available data on corrosion needs to be compiled and monitored on continuous basis.

CONSENT FOR PUBLICATION

Not applicable.

CONFLICT OF INTEREST

The author declares no conflict of interest, financial or otherwise.

ACKNOWLEDGEMENTS

Declared none.

REFERENCES

[1]   Best Available Techniques for Pollution Prevention and Control in the European Fertilizer Industry. *Production of urea and urea ammonium nitrate*; EFMA, **2000**.

[2]   Juneja, D.; Kumar, J. Fabrication of Construction Materials in Urea Manufacturing Plants. *International Journal of Enhanced Research in Science Technology & Engineering,* **2013**, *2*(9), 56-59.

[3]   Shaikh, H.; Subba Rao, R.V.; George, R.P.; Anita, T.; Khatak, H.S. Corrosion failures of AISI type 304 stainless steel in a fertiliser plant. *Eng. Fail. Anal.,* **2003**, *10*(3), 329-339.
        [http://dx.doi.org/10.1016/S1350-6307(02)00076-6]

[4]   Eker, B.; Yuksel. E. Solutions to corrosion caused by agricultural chemicals, trakia. *J. Sci.,* **2005**, *3*(7), 1-6.

[5]   Craig Meyers. "Take UAN Corrosion seriously. *Fluid J.,* **2009**, *7*(1), 63.

[6]   Sharma, P. Microbiological-influenced corrosion failure of a heat exchanger tube of a fertilizer plant. *J. Fail. Anal. Prev.,* **2014**, *14*(3), 314-317.
        [http://dx.doi.org/10.1007/s11668-014-9826-2]

[7]   Uba, B.N. Microbiological characteristics of wastewaters from a nitrogen- and phosphate-based fertilizer factory. *Bioresour. Technol.,* **1995**, *51*(2-3), 143-152.
[http://dx.doi.org/10.1016/0960-8524(94)00105-A]

[8]   Ahlgren, S.; Baky, A.; Bernesson, S.; Nordberg, K.; Norén, O.; Hansson, P-A. Ammonium nitrate fertiliser production based on biomass - environmental effects from a life cycle perspective. *Bioresour. Technol.,* **2008**, *99*(17), 8034-8041.
[http://dx.doi.org/10.1016/j.biortech.2008.03.041] [PMID: 18440225]

[9]   Anonymous, *Corrosion control of agricultural equipment and buildings,* http://www.npl.co.uk/ncs/docs

[10]   Kar, S.C.; Rajan, R.G. Improving reliability of fertiliser plants. *Indian J. Fert.,* **2010**, *6*(6), 12-16.

[11]   Guy Schneider & Olivia Woerth, The Use of FRP. *(Fiberglass-Reinforced Plastic) in Phosphate Fertilizer and Sulfuric Acid Processes, Ashland Performance Materials, paper no. 12*; CORCON, **2017**.

[12]   Inspection Manual for Fertilizers Industry. **2012**. www.eeaa.gov.eg/ippg /.../Fertilizers /Fertilizers%20english/Sec%202.doc

[13]   Üneri, S. *Corrosion and prevention*; Corrosion Foundation Publication: Ankara, Turkey, **1998**.

[14]   Ürgen, M. Corrosion on Stainless Steel *Corrosion science and Techniques Journal, Ankara, Turkey,* **1989**.

[15]   Akdogan, A.; Eker, B. The effect of corrosion on pump system and prevention methods, 19. *Agricultural Mechanization Congress,* Erzurum, Turkey**2000**.

[16]   Akdogan, A.; Eker, B. The effect of corrosion on agricultural machinery and prevention methods *Denizli Material Congress,* 26-28 April 2000Denizli, Turkey**2000**.

[17]   Anonymous, *Corrosion control of agricultural equipment and buildings,* http://www.npl.co.uk/ncs/docs

[18]   Özkan, E. *Spraying, Equipment: Pumps*; Iowa State University: USA, **1989**.

[19]   Saha, A. *Boiler tube failures. Handbook of Materials Failure Analysis with Case Studies from the Chemicals*; Concrete and Power Industries, **2016**, pp. 49-68.

[20]   Sastry, V P; Manian, C V Corrosion – tacking challenge in indian fertiliser industry, technical committee, NACE international gateway india section, **2013**.

CHAPTER 12

Marine Corrosion

H. Manjunatha[1,*], K. Venkata Ratnam[1], S. Janardan[1], R. Venkata Nadh[1], N. Suresh Kumar[2], N.V. Krishna Prasad[3], S. Ramesh[3], K. Chandra Babu Naidu[3] and T. Anil Babu[3]

[1] *Department of Chemistry, GITAM Deemed to be University, Bangalore - 562163, Karnataka, India*

[2] *Department of Physics, JNTU College of Engineering Anantapur, Anantapuramu-515002, A.P., India*

[3] *Department of Physics, GITAM Deemed to be University, Bangalore-562163, Karnataka, India*

Abstract: Seawater is a hostile environment not only for people but also for metals and alloys. It is often considered that sea water is the most severe environment to which materials can be exposed. Warmer water accelerates the rate of corrosion due to high temperature and particularly aggressive. In this chapter, marine corrosion, its mechanism, factors affecting corrosion and several methods adopted for the prevention of corrosion are described with an emphasis on marine corrosion inhibitors. Organic compounds containing hetero atoms like O, N, S, *etc*., along with double and triple bonds are found to be very effective in preventing marine corrosion of alloys and metals. Most of the corrosion inhibitors are found to show inhibition property by getting adsorbed on to the metal surface through the principle of different adsorption isotherms known. The maximum corrosion inhibition efficiency of organic inhibitors is found to be more than 99%. The use of inorganic compounds and paints as corrosion inhibitors is discussed.

Keywords: Corrosion Inhibitors, Factors Affecting Corrosion, Marine Corrosion, Mechanism of Corrosion, Stainless Steel & Alloys.

1. INTRODUCTION

Corrosion is a 'billion-dollar thief'. Even though it is a natural phenomenon, it results in loss of material, money and life. Metals have a strong crystalline structure, and due to corrosion, they get converted into their salts, making them losing their metallic strength resulting in damage to machinery, structures or equipment in which they are used. Thus, corrosion causes damage to metals and

---

* **Corresponding author H. Manjunatha:** Department of Chemistry, GITAM Deemed to be University, Bangalore-562163, Karnataka, India; Tel: +91- 88611 58964; E-mail: hanumanjunath80@gmail.com

N. Suresh Kumar, P. Banerjee, H. Manjunatha and K. Chandra Babu Naidu (Eds.)

thereby to society. Corrosion is considered as a most damaging and provocative menace in many countries including India causing approximately a loss of one to five percent in Gross National Product (GNP) of each country as per estimations of NACE (National Association of Corrosion Engineers) [1]. A recent investigation by NACE estimated annual global cost of corrosion to be approximately US $2.5 trillion, which is almost equal to 3.4% of the global GDP [2, 3]. In the US, the total cost of corrosion is more than the US $1.1 trillion [4] annually. The annual corrosion cost of India is more than US $100-billion and that of South Africa; the direct corrosion cost is estimated to be around the US $9.6 billion [2, 3] which may be increased with technology development and usage of metallic substances. This loss of money all over the globe due to corrosion can be reduced up to 35% by applying proper prevention methods. Corrosion also results in loss apart from large economic damage.

For example, an explosion due to corrosion of metallic structure killed over 200 people in Guadalajara, Mexico, in April 1992 [5]. In addition, these fatalities, the series of blasts damaged 1,600 buildings and 1,500 people injured and the total loss due to both direct and indirect effects of corrosion is estimated to be 75 million US dollars [6]. The cost of corrosion is more than the total annual cost of floods, hurricanes, fires, lightning, earthquake and other natural calamities that occur all over the globe annually. Hence, attention is to be given by corrosion, and similarly, the substantial measures may be taken in reducing economic loss and loss of life.

2. WHY DO METALS UNDERGO CORROSION AND WHAT IS THE DRIVING FORCE?

As per Roberge, "A material destructive attack on reacting with the environment is Corrosion" [7]. In general, it is a known fact that most of the metals are unstable in their free state. Extraction of metals from their combined states (a process called Metallurgy) involves the addition of a large amount of heat energy. The metals (Free State) extracted from their ores will be at higher energy state and have a tendency to go into their combined form generally the corrosion products like rust and scale [8, 9]. Thus, they have a natural tendency to revert back to their combined form when exposed to attacking the environment. Thermodynamically, the standard free energy change (ΔG^0) decides the relative rate of corrosion of metals, and in general, higher the ΔG^0 value, greater will be the corroding tendency and *vice versa* [10, 11].

Metal Ore   Metallurgical operations →   **Pure Metal**   Action of environment →   **Corroded Metal**
(thermodynamically   + Energy   (thermodynamically   (comparatively more
stable state)   unstable state)   stable than pure metal)

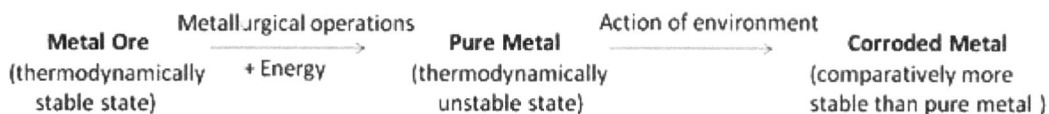

In this chapter, we restrict ourselves to marine corrosion, its mechanism, causes, factors affecting marine corrosion and the latest developments in the various ways of preventing it.

3. MARINE CORROSION AND ITS MECHANISM

Seawater is an aggressive environment for people, metals and alloys. It is considered often that seawater is the most severe environment to which materials can be exposed. Hot waters speed up the corrosion rate due to high temperature and particularly aggressive. This was the major problem towards the end of the Second World War, which led to Pacific campaigns when military equipment got corroded at an unexpected rate in the tropics [12]. Also, corrosion may be severe in colder waters. Various types of destruction can occur to ships, structures and other equipment used in seawater services. Marine Corrosion is a term that describes many problems faced by marine equipment of metal base used in production, processing and transport industries such as pipelines, households, fuel tanks, oil lines, power lines, heat conductors, heat exchangers and several other marine applications. Marine platforms such as ships, offshore structures and steel bridges are at higher risk of corrosion exposed to the seawater environment, particularly after fifteen to twenty years. Cargoes on-board oil tankers and bulk carriers may be highly corrosive. Incidents like Erika in 1999 [13] and Castor in 2000 [14] highlight the threats of corrosion for ships. Fig. (**1**) displays an image of severe corrosion under deck area of an oil tanker [15]. As corrosion may lead to structural material degradation, facilitate fatigue cracks, brittle fracture and unstable failure [16], the integrity of the entire hull structure can be affected considerably.

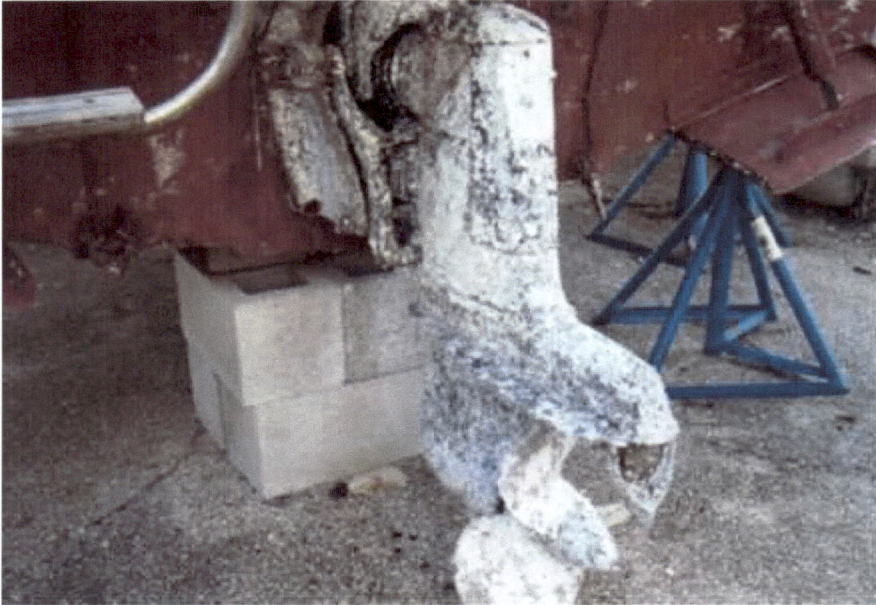

Fig. (1). Marine Corrosion.

4. CORROSION MECHANISM IN MARINE ENVIRONMENTS

Corrosion nature and its mechanism in seawater environment depend on the metal and alloys used. Generally used materials in marine environments are steel, stainless steel, aluminium, copper-nickels and reinforced concrete [17]. Among these, iron or steel is the material used extensively for marine structures, although aluminium and other metals and alloys are used for many applications. Thus, the corrosion of iron or steel follows the below mechanism. According to the electrochemical theory of corrosion, the surface of iron contains anodic (iron atoms) and cathodic sites (impure atoms like carbon). At anodic sites, iron oxidizes to give Fe^{2+} and Fe^{3+} (ferrous and ferric) ions (Eq.1) with the release of electrons [18]. These ions and chloride ions combine to give median species as per equations (3) and (4) [18, 19].

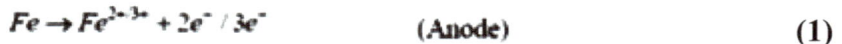

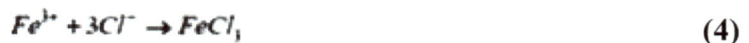

$$Fe \rightarrow Fe^{2+,3+} + 2e^- / 3e^- \qquad \text{(Anode)} \qquad (1)$$

$$O_2 + 2H_2O + 4e^- \rightarrow 4OH^- \qquad \text{(Cathode)} \qquad (2)$$

$$Fe^{2+} + 2Cl^- \rightarrow FeCl_2 \qquad (3)$$

$$Fe^{3+} + 3Cl^- \rightarrow FeCl_3 \qquad (4)$$

Later these median species combine with moisture or water forming hydrochloric acid, ferrous [Fe $(OH)_2$] and ferric [$Fe(OH)_3$] hydroxides (Eq. 5 & 6).

$$FeCl_2 + 2H_2O \rightarrow Fe(OH)_2 + 2HCl \tag{5}$$

$$FeCl_3 + 3H_2O \rightarrow Fe(OH)_3 + 3HCl \tag{6}$$

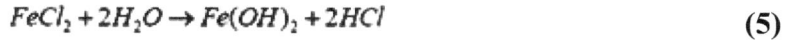

These hydroxides can also be formed by combining ions of hydroxide created at cathode sites due to reaction between moisture, oxygen and electrons released from the anodic sites according to the equations 2, 7 & 8.

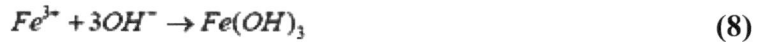

$$Fe^{2+} + 2OH^- \rightarrow Fe(OH)_2 \tag{7}$$

$$Fe^{3+} + 3OH^- \rightarrow Fe(OH)_3 \tag{8}$$

The iron hydroxides formed in the previous steps further reacts with oxygen and moisture to form the final corrosion product, *i.e.* rust.

$$4Fe(OH)_2 + O_2 + x\,H_2O \rightarrow 2[Fe_2O_3 \cdot xH_2O]$$
$$\text{Rust}$$

Sodium chloride present in seawater enhances the iron corrosion rate by moulding the intermediate species and hydrochloric acid formation in accordance with equations (5 and 6) which decreases the pH sea water and make it more aggressive medium towards corrosion.

Aluminium undergoes corrosion in seawater through the formation of several intermediate species depending on the pH of the environment. In the presence of sodium chloride in seawater, aluminium undergoes oxidation into its ion rapidly at the anode ($< 10^{-6}$s).

$$Al \rightarrow Al^{3+} + 3e^- \qquad \text{(Anode)} \tag{9}$$

$$O_2 + 2H_2O + 4e^- \rightarrow 4OH^- \qquad \text{(Cathode)} \tag{10}$$

The Al^{+3} ions thus formed hydrate quickly and hexa-coordinated complex is formed in a microsecond. Hydrolysis of Al^{+3} results in the formation of a variety of mononuclear hydrolysis species according to the following equations [20, 21].

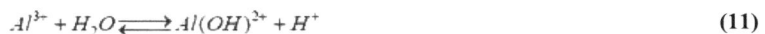

$$Al^{3+} + H_2O \rightleftharpoons Al(OH)^{2+} + H^+ \tag{11}$$

$$Al^{3+} + 2H_2O \rightleftharpoons Al(OH)_2^+ + 2H^+ \tag{12}$$

$$Al^{3+} + 3H_2O \rightleftharpoons Al(OH)_3 + 3H^+ \tag{13}$$

$$Al^{3+} + 4H_2O \rightleftharpoons Al(OH)_4^- + 4H^+ \tag{14}$$

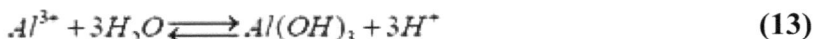

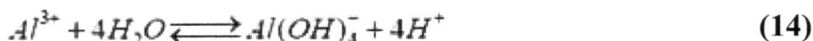

Poly-nuclear hydrolysis product formation has been postulated [22, 23] for aluminium in addition to the above mentioned mononuclear hydrolysis products.

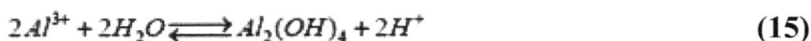

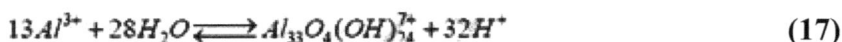

$$2Al^{3+} + 2H_2O \rightleftharpoons Al_2(OH)_4 + 2H^+ \tag{15}$$

$$3Al^{3+} + 4H_2O \rightleftharpoons Al_3(OH)_4^{5+} + 4H^+ \tag{16}$$

$$13Al^{3+} + 28H_2O \rightleftharpoons Al_{33}O_4(OH)_{24}^{7+} + 32H^+ \tag{17}$$

Otherwise, the formation of the intermediate species in the presence of chloride ions itself acts as products of corrosion or even converted to different chemical species.

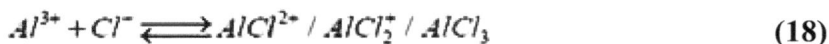

$$Al^{3+} + Cl^- \rightleftharpoons AlCl^{2+} / AlCl_2^- / AlCl_3 \tag{18}$$

The chemical species formed as corrosion product are highly reactive and soluble in the corrosive media like H_2O, HCl and NaCl, *etc.* Thus, the presence of chloride ions in seawater enhances the rate of corrosion. Copper is another important metallic material used in the marine industry or environment because of high electrical and thermal conductivities, high class chemical properties and excellent working. Copper metal is known to be best in power lines, heat exchanges, heat conductors, domestic water pipelines and pipelines for sea water [22, 23]. Copper exhibits the tendency of getting corroded on exposure to seawater containing NaCl analogous to aluminium and steel. Moisture enhances the cathodic reaction rate of copper corrosion as per the equation.

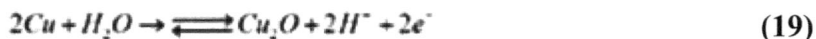

$$2Cu + H_2O \rightarrow \rightleftharpoons Cu_2O + 2H^- + 2e^- \tag{19}$$

The literature also shows that oxygen also initiates corrosion of copper and it propagates in two steps as follows [23, 25]:

$$2Cu + \frac{1}{2}O_2 + 2H^+ \longrightarrow 2Cu^+ + H_2O \tag{20}$$

$$2Cu + \frac{1}{2}O_2 + 2H^{\cdot} \longrightarrow 2Cu^{\cdot} + H_2O \qquad (21)$$

The cuprous (Cu^+) and cupric (Cu^{2+}) ions can also form CuCl and $CuCl_2$ respectively upon reacting with NaCl of sea water. Thus, CuCl, $CuCl_2$ and Cu_2O become the major corrosion products formed. In addition, in literature, it is postulated that a $Cu_2(OH)_3Cl$ (highly soluble atacamite) is formed as a product of corrosion of copper according to the following equation [22, 24]:

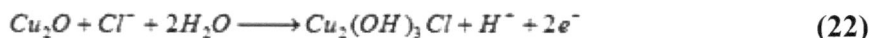

$$Cu_2O + Cl^- + 2H_2O \longrightarrow Cu_2(OH)_3Cl + H^{\cdot} + 2e^- \qquad (22)$$

5. FACTORS AFFECTING MARINE CORROSION

Directly or indirectly lot of variables in the ocean water can influence the marine corrosion, which was referred by various articles [24 - 28] and some of the important ones are highlighted.

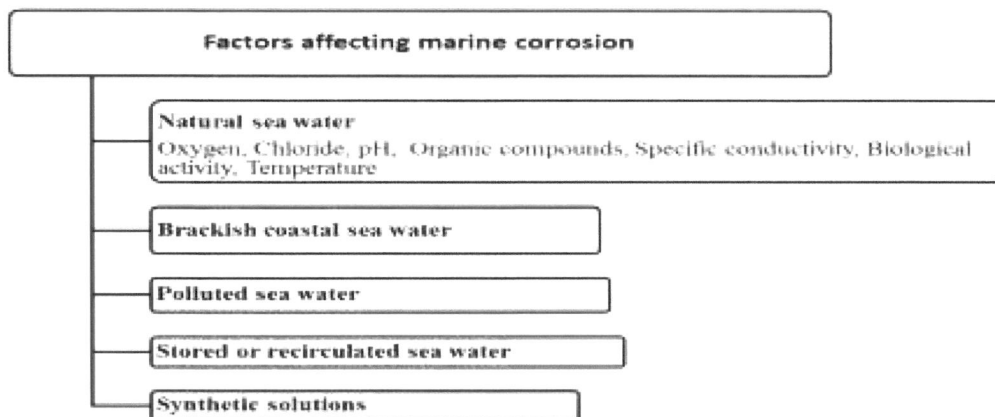

```
┌─────────────────────────────────────────────────────────────┐
│          Factors affecting marine corrosion                  │
└─────────────────────────────────────────────────────────────┘
    │
    │   ┌───────────────────────────────────────────────────────┐
    ├───│ Natural sea water                                     │
    │   │ Oxygen, Chloride, pH, Organic compounds, Specific     │
    │   │ conductivity, Biological activity, Temperature        │
    │   └───────────────────────────────────────────────────────┘
    │   ┌───────────────────────────────────┐
    ├───│ Brackish coastal sea water        │
    │   └───────────────────────────────────┘
    │   ┌───────────────────────────────────┐
    ├───│ Polluted sea water                │
    │   └───────────────────────────────────┘
    │   ┌───────────────────────────────────┐
    ├───│ Stored or recirculated sea water  │
    │   └───────────────────────────────────┘
    │   ┌───────────────────────────────────┐
    └───│ Synthetic solutions               │
        └───────────────────────────────────┘
```

Metals and nonmetals are destroyed not only due to ocean water but also due to other assisting factors associated with seawater such as microorganisms, weed clustering, silt and slime which are, despite not having oxygen usually creates severe corrosive environment under the deposits. Moreover, the coatings and composite structures are highly exposed to rapid corrosion. In addition, sulphate bacteria, which is present in the silt or mud, enhances the concentration of hydrogen sulphide, which is particularly aggressive to steel and copper-based alloys.

6. NATURE OF SEA WATER

6.1. Oxygen

High pH conditions of the ocean water make the dissolved oxygen as the only oxidation system, and it is true for the majority of the metals except for Mg which shows high negative standard potential and may reduce the water also. However, this general rule has some exceptions such as reduction of the protons and ions in occluded or restricted zones such as crevices, cracks and pits consisting of variable components when compared with the normal ocean water outside these zones. In addition to the oxygen present in ocean water, organic matter as well as microorganism (bacteria), triggers the corrosion. The reduced form of oxygen always plays a major role in metal degradation process and depends upon the percentage of oxygen.

The oxygen content will influence corrosion potential and corrosion rate in case of actively corroding metals. In numerous corrosion processes, the cathodic reaction will be under the control of mass transfer and flow speed that determine the rate. The requirement of oxygen in keeping and fixing the inactive film is necessary for dynamic detached metals. In addition, the dissolved oxygen influences corrosion types like pitting and hole corrosion. In this regard, it was demonstrated that pit formation was inhibited by low dissolved oxygen concentration and development of pit decreases on aluminium composite AA 5052 in ocean water

6.2. Chloride

High Chloride content is found to be significant for the development of metallic complexes (*e.g.*: Cu, Fe) and accelerates the rate of corrosion reaction both qualitatively as well as quantitatively. The effect of high chloride concentration is one of the factors that lower metals redox potential and extend possible corrosion reaction. The increasing rate of oxidation in metal is another effect observed when complex intermediates are formed. For instance, as indicated by certain creators, the chloride particle is legitimately engaged with the oxidation system [29]. Moreover, the mechanistic aspects of metal ion reaction with seawater are still rather incomplete [30 - 32].

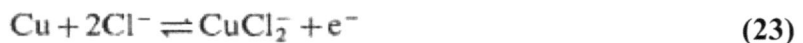

$$Cu + 2Cl^- \rightleftharpoons CuCl_2^- + e^- \tag{23}$$

6.3. pH

In general, alloys utilized in ocean water frameworks are not affected with 7.8-8.3 pH range (normal). Nevertheless, some applications require a change in the composition of water, for example, desalination of water through reverse osmosis in the ocean. Due to a reduction in the pH of incoming water into the corrosive area, deposition of calcium carbonate is prevented and as a result, reduces the membrane blockage. The sea water thus acidified has significant corrosive action towards composites, which must be considered. Similarly, a decrease in pH value to 7.2 uniquely expands the susceptibility of aluminium-magnesium composites to restricted corrosion. The propagation rate increases once the initiation starts [33].

6.4. Calcium and Magnesium

The higher pH values of natural ocean water reduce the dissolved oxygen when it is in contact with the surface of metals and enhances the formation of calcareous layers by inhibiting the formation of calcium oxides, magnesium oxides, salts and hydroxides [34]. Calcareous layers inhibit the interaction of oxygen through cathodic protection by decreasing the current demand [30]. Species like brass, aluminium, magnesium and calcium and magnesium have been very important during the ion exchange process in forming a protective layer with sea water [35]

6.5. Specific Conductivity

The general and localized corrosions depend on sea water's specific conductivity on a relative basis. The influence of specific conductivity is more in general corrosion, and it is even more pronounced in case of localized corrosion due to the fact that real current flow during the corrosion is more. This effect could become also reversed in some cases. For instance, in bimetallic corrosion, the enhancement of the surface area of the metal enhances the corrosion behaviour due to high conductivity [36]. In addition, we can see some reduced rate of corrosion with the water with lesser conductivity.

6.6. Biological Activity

Corrosion can be affected by fouling [37, 38],

○ Protecting metal surface from oxygen supply help information of differential aeration cells among protected and unprotected areas.

○ We involve animal secretion in the process of corrosion.

○ Catalytic effects.

○ Confining the water flow depending on the situation.

○ Degrading fouling can deliver sulfides that modify the environment on the metal surface.

Microscopic bacteria can have an articulated impact [39 - 41] given by:

○ It is controlling corrosion logically or in a roundabout way by affecting the redox component and thereby decreasing its concentration on the metal surface.

○ It is modifying redox reactions that occur at the metal surface either through hindering or catalyzing incomplete corrosion reactions [42 - 44].

6.7. Temperature

The rate of corrosion of steel nearly doubles for every 10° C rise in temperature. Temperature affects mass transfer and kinetics of corrosion reaction rate. Apart from its immediate effect on the rate of corrosion, the temperature can also modify the properties of corrosive products [45, 46]. Apart from this, bacterial slimes are temperature dependant, which influences free corrosion in oceans.

7. BRACKISH SEA WATER

Saline water composition varies in natural oceans depending on the local environments and thus in the manner in which it influences marine corrosion. However, the important differences among saline waters are as follows:

i. The decrease in the concentration of salt changes oxygen content, enhancing temperature and contamination. Diminished oxygen levels hinder the pace of general corrosion.

ii. Decreased chloride content because of enhanced dilution prevents the formation of complex properties, raising the potential for pitting and crevice erosion.

iii. Enhanced dilution decreases specific conductivity reducing the impact of corrosion.

iv. Major constituents of ocean water stay the same if there should be an occurrence of common or saline water. However, there will be an extraordinary decrease in the concentrations (few orders of magnitude) of minor constituents in saline water because of dilution. This significantly influences the formation of corrosion product which depends upon minor constituent's concentration.

v. The diversity of organic compounds with its concentration gradually enhances the complexation with the metal ions and rate of corrosion.

vi. Enhanced fouling frequently arises in saline waters prompt increase in shielding effects, thereby reducing general corrosion rate owing to the reduction in oxygen levels. However, corrosion may increase due to increased bacteriological activity underneath the surface of the fouled metal.

vii. Higher concentrations of suspended solids in brackish waters have an impact on the corrosion processes, often coupled with water velocities. At low velocities, solids deposit on metal surfaces which leads to the risk of corrosion due to the formation of differential aeration cells, deposit attack, *etc.* [47].

7.1. Polluted Sea Water

Polluted sea water is characterized by low oxygen content, reduced pH and presence of sulphide and /or ammonia ions [48], which encourage unique types of corrosion mechanism that generate different layers of corrosion product contrary to naturally formed in marine water. It is identified that introduction of marine metals to alternative polluted and natural sea waters is extremely detrimental [49, 50].

A number of copper alloys are affected in particular by these high sulphide levels and become more prone to pitting. This type of pollution is more enhanced when bio-fouling present in systems which upon decaying produces sulphur-based compounds. This is the major form of pollution problems, particularly when the use of hypochlorite and other biocides are prohibited. In a huge cooling system, this process can play a significant role as a result of the production of biological matter in the form of mussels, barnacles and shellfish as thick layers. Moreover, in stagnant water bodies which are more anaerobic and resulting in the death of organisms which decomposed gradually, this effect is more enhanced. During stagnant or low flow water conditions, the system is likely to become anaerobic, resulting in the death of the organisms followed by their gradual decomposition.

7.2. Synthetic Solutions

Most of the synthetic solutions are free from organic, biological, and bacteriological species. The mechanism of corrosion always varies when compared with natural ocean water. However, in the majority of the synthetic solutions, for example, 3.5% NaCl, only the chloride effect is taken, although bicarbonate ions are sometimes added to imitate ocean water.

7.3. Types of Marine Corrosion

The following are the predominant types of corrosion that occur in seawater are (Fig. **2**):

○ General corrosion

○ Galvanic corrosion

○ Pitting

○ Crevice corrosion

○ Exfoliation corrosion

○ Erosion corrosion and cavitation

○ Stray current corrosion

○ Stress corrosion cracking

○ Corrosion fatigue

Fig. (2). Important type of marine corrosion.

7.4. General Corrosion

Corrosion attack that progresses uniformly on the complete metal surface is known as general corrosion. Examples include plain carbon steel that corrodes in ventilated seawater, metal or alloy becoming progressively thin due to corrosion until its thickness reduces to a point susceptible to failure. If carbon and low-alloy steels are exposed to the seawater environment, the typical corrosion rates do not vary or vary from steel to steel. When carbon and low alloy steels are continuously exposed to seawater, it should be noted that they are cathodically preserved. In the case of a ship hull, the steels are painted with an anti-corrosion paint, stain-resistant paint and then they are cathodically secured.

7.5. Galvanic Corrosion

When two different metals are joined together and exposed to an electrolyte, they form a galvanic cell. The anodic metal is oxidized and corroded. For example, zinc and copper metals bonded to each other in an electrolyte medium form a galvanic cell. Zinc acts as an anode and is subject to corrosion while being cathode, copper is not affected. This type of electrochemical corrosion is also called bimetallic corrosion. The potential difference between the two metals is the cause of corrosion. If two metals are submerged in seawater, each develops its own electrode potential core such that noble E_{corr} value denotes cathodic metal highE_{corr}value denote anodic metal. The flow of current between cathode and anode can be measured. Galvanic effects are more important in marine hardware systems made of different alloys. It is important to select noblest alloys for the core components in marine systems. Measurement and comparison of E_{corr}can estimate formation of galvanic corrosion in sea water by two different metals. If the values vary by hundreds of milli-volts, clear galvanic corrosion is likely to occur while welding two alloys with small differences in E_{corr} values may lead to increased corrosion.

7.6. Pitting

The name pitting implies a form of localized corrosion attack on metals and alloys which develop pits. When a metal or alloy has a protective oxide layer against corrosion and if that protective oxide film breaks down locally due to chemical attack then it leads to pitting corrosion. Common causes for the onset and spread of pitting corrosion according to research are [51]:

○ Localized chemical or mechanical damage to the protective oxide film.

o Factors contributing to the breakdown of inert protective films such as acidity, low dissolved oxygen concentrations and high chloride concentrations. These can make a protective oxide film less stable, thereby starting the pit.

o Localized damage or poor use of protective coating.

o Absence of homogeneity in the metal structure of components such as non-metallic additions.

Corrosion in engineering structures can be devastating in the drilling of tools,creating stress in complex areas. At the same time toleration of non-perforated shallow surface pit is sometimes accepted for economic reasons. Particularly, alloys of aluminium are vulnerable for pitting corrosion in seawater while stainless steels rarely form pits on open surfaces at sea and when excavated, they are usually associated with deposits or contaminants that form cracks and are called crevice corrosion.

7.7. Crevice Corrosion

Crevice corrosion is a form of localized corrosion that occurs due to the presence of a stagnant solution on cracks or shielded surfaces. Metal/metal or non-metal/non-metal junctions like bolts, rivets, gaskets and surface deposits are examples of such cracks or shielded surfaces. Fouling, especially hard-shell fouling could also be the cause for seawater crevice corrosion at ambient temperature. The mechanism for crevice corrosion in chloride solution can be analyzed in terms of chloride-ion concentration, oxygen depletion,metal-build up sequences, hydrolysis of metal ions leading to local acidification, protective film rupture and lateralization.

7.8. Exfoliation Corrosion

This type of attack, also called exfoliation corrosion, occurs in aluminium alloy series such as 5000 and 7000. The corroded plate contains alternate plates of corroded and un-corroded metal. Exfoliation corrosion is found to occur in 5456 alloys in tempered H343, in 5086 alloys in tempered H34 and 7178, 7079, 7075, 7039 alloys in tempered T6. Exfoliation corrosion is inter-granular in these alloys and is related to the heat treatment process which precipitates the corrodible inter-metallic particles that meet at the alloy boundary of the grain embedded in the coating side. Exfoliation corrosion can be trans-granular in Al-Zn-Mg-Mn alloys. Under harsh service conditions such as those involving moisture and chloride, it is necessary to protect al-Zn-Mg-Mn alloys. Products that are in a hot state will be at risk of being fired and as a result, may need full protection. Both metal binding and stainless-steel alloy installation have been shown to be very effective in providing that protection.

7.9. Erosion-Corrosion and Cavitation

Attack due to acceleration of fast-flowing sea water consisting of solid particles sometimes capable of cutting or wear is known as Erosion-Corrosion. These fluids are capable of removing protective films on the metal surface and are more aggressive if the frictional force of the fluid is high at high speeds. In general, the higher the speed, more polished is the solution. Since the effect of corrosion is more on the mechanical cutting process, the resistance of this attack is known by the balance of properties enhance corrosion resistance and enhance resistance to erosion. Accelerating factors are turbulent and patterns that cause obstruction to change its direction of flow. Remedies include a change in design, usage of high resistant compounds, protection of cathode and removal of suspended solids.

Damage of cavitation is another attack under the condition where the high relative velocity exists between solids and liquids. For example, it occurs in pumps, hydraulic turbines and ship propellers based on the theory of hydro-mechanics. In brief, the fluid that comes under flow divergence, rotation, or vibration forms low-pressure areas that produce cavities or bubbles. The formed cavities collapse at extreme pressures producing intense shock waves. Observation of cavitation damage on glass and plastic confirmed that this phenomenon is primarily a process of corrosion. However, corrosion accelerates damage cavitation in alloys and metals in a corrosive environment and sometimes may cause attack under mild conditions of cavitation rupturing only protective surface films. Alloys like Inconel 625, Hastelloy C-276 and molybdenum contained austenitic stainless steel are found to be highly resistant to cavitation damage.

7.10. Stray-Current Corrosion

This type of corrosion is an outcome of direct current flow from seawater to the ship's hull. When the DC current flows through the conductive solution like sea water, it causes the corrosion of most of the metals in the sea water and its magnitude of damage to the metal structures depends on the quantity of current discharged. In the majority of cases, the stray-current corrosion occurs at the point or place where the current enters the sea water from its source. Nevertheless, in the case of zinc and aluminium, this corrosion occurs at places where stray currents are picked up from sea water attributed to excessive alkaline conditions. Aluminium alloys have more tendency towards stray-current corrosion which is more rapid on coated structures than on bare structures.

7.11. Stress Corrosion Cracking

Failure due to spreading of cracks in corrosion environment is known as stress corrosion cracking. This is due to the presence of tensile stress provided loads, welding, heat treatment, *etc*. Cracks are formed and spread at right angles to the stress towards much lower stress area. When the stress crack propagates to a depth where the remaining phase of the alloy load reaches air pressure, the content is separated by excessive normal cracks. Failure, when it happens, is therefore often dangerous and that is why so much attention is paid. Different alloys that show this corrosion in seawater include high-strength titanium,s teel, titanium alloy with fracture cracking and many. Stress corrosion cracking for high-strength alloys has a resistance index, Klscc [19] and is defined as stress intensity below which stress corrosion cracking is not expected to occur.

7.12. Corrosion Fatigue

Cracking of metal under repeated cyclic stresses is known as corrosion fatigue which develops and propagate stress below the produced stress after many cyclic applications of stress. Similar to stress corrosion cracking, fatigue crack propagates until the load-bearing section of the material reduces to a point at which ultimate strength is exceeded, and the material separates by overload fracture. In the evaluation of alloys corrosion fatigue resistance with respect to their stress at which no failure occurs is above 10 or 108 cycles. This stress is known as corrosion fatigue strength (CFS). It is emphasized that laboratory test data described in Reference 18 should not be used directly to predict the service life of components, such as ships propellers. In general, both the time of exposure the particular material to the corrosive environment and the frequency and magnitude of the stress cycles are significantly different for a material in service and should be taken into consideration. For this reason, engineers often study the rate at which corrosion fatigue cracks grow in order to be able to predict the remaining life of cracked structural components.

8. MARINE CORROSION PREVENTION

8.1. Cathodic Protection

Metals normally undergo corrosion by an electrochemical process with the formation of anodic and cathodic regions in contact with each other. Corrosion of metal takes place at the anodic region whereas at the cathodic region, the metal is unaffected. Therefore, corrosion can be prevented by eliminating the anodic sites and converting the entire metal into the cathodic area. There are two methods:

○ Impressed voltage method.

○ A sacrificial anodic protection method.

8.1.1. Impressed Voltage Method (ICCP)

The impressed voltage or ICCP system generate electrons from an external DC power source. The ICCP system contains a rectifier, anode, reference electrode and a controlling unit [52, 53]. The ICCP system provides remarkable protection against all types of marine corrosion of ships and offshore platforms, pipelines, ports, and steel piles, *etc.* and there it is widely used in marine corrosion protection. However, ICCP system, despite having several advantages, has certain drawbacks [54] and are as follows:

○ The requirement of skilled workers.

○ A continuous power supply must be sustained.

○ The current must always be connected in the right direction.

○ If permanent anodes are used, then-current shields are required.

8.1.2. The Sacrificial Anodes Cathodic Protection System (SACP)

The ICCP method discussed earlier is not desirable for older ships which are larger in number with less lifespan due to its high initial installation cost. In such cases, cost-effective, SACP system is more preferred due to its low cost for a short period. The basic working principle of SACP is that attaching a metal which readily undergoes corrosion compared to the entire marine structure (Steel) and converting the marine structure like a ship's hull into a cathodic region. The potential difference between the steel to be protected and a second metal in the same environment causes the driving voltage. Anode undergoes corrosion protecting the cathode (marine structure). A galvanic cell is set in due to difference in electrode potentials of the two metals redox reaction can then occur between the two metals, with electron transfer occurring from the anode to the cathode driven by their difference in electrochemical potentials. The large gap or difference in electrochemical potentials between the sacrificial anode and the ship's hull avoids corrosion occurring on the hull surface itself (cathodic region, No oxidation, thus no corrosion) and instead, corrosion occurs only on the attached anode and thus it is given the term "sacrificial anode" [55, 52]. Generally used anodes are aluminium, zinc or magnesium alloys which are anodic with respect to steel materials [56]. Finding a suitable anode is an important factor. It is recommended that 15-20% of the sacrificial anodes should be installed to the stern and rudder area of the ship [54]. Following are some of the advantages and disadvantages of SACP. Sample ICCP and SACP is shown in Fig. (3).

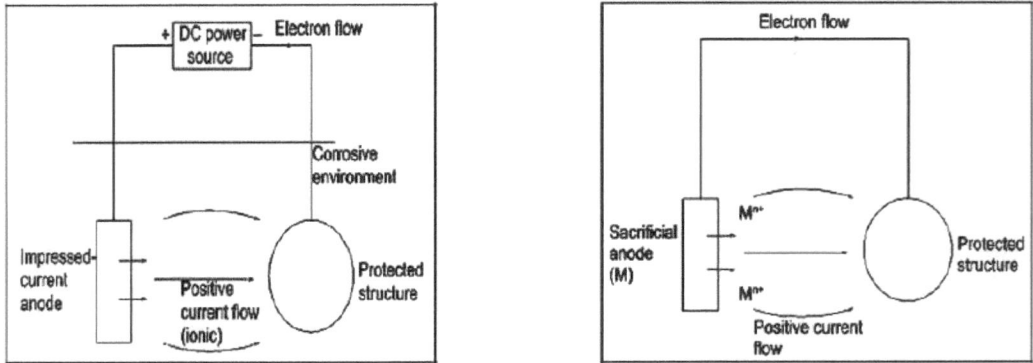

Fig. (3). A simple ICCP & SSCP models.

○ No power supply to be installed.

○ Very simple to maintain and use.

○ More expensive than an ICCP system for long term operation, although it has low initial costs.

9. CORROSION INHIBITORS

9.1. Inorganic Salts as Inhibitors

Most effective compounds used as corrosion inhibitors for different metals and alloys [57, 58] are chromates. However, increasing environmental concern and strict environmental laws never permit the use of chromate-based corrosion inhibitors due to their high toxicity and negative impact on the environment [57, 58]. Therefore, much of the work has been devoted to developing corrosion inhibitors that are environmentally benign in nature. In this scenario, many inorganic salts either in a free state or combined with organic compounds have been investigated as promising corrosion inhibitors [59, 60]. Aramaki [63, 64] explored the corrosion inhibition effect of Al^{3+} (AlCl3), La^{3+}(LaCl$_3$), Ce^{3+} (CeCl$_3$) and Ce^{4+}(Ce (SO$_4$)$_2$) on zinc in 0.5M NaCl solution using X-ray photoelectron spectra (XPS) and potentiodynamic polarization methods. The study found that

$La^{3+}(LaCl_3)$ and $Ce^{3+}(CeCl_3)$ inhibit the zinc corrosion rate by more than 90% efficiency whereas Al^{3+} and Ce^{4+} enhanced the zinc corrosion rate in the test medium [58]. Apart from the above inorganic compounds, few lanthanide ionic salts have also been explored as corrosion inhibitors for iron and its alloys [61, 62], aluminium and its alloys [58, 64 - 71] and zinc [72 - 74]. The effect of nickel concentration on the corrosion behaviour of Cu-Ni alloy was also studied by Badawi *et al*. [75], in 0.6 M NaCl solution using electrochemical impedance spectroscopy (EIS) and showed that higher concentration of Ni significantly reduced the corrosion rate.

9.2. Plant Extracts as Inhibitors for NaCl Media

Green corrosion inhibitors quote plant extracts as much attractive in view of their nature. Often, these green corrosion inhibitors are used for protecting various metals and their alloys in a wide range of electrolytic media, including ocean water [76 - 85]. The rules behind the determination of plant extracts as inhibitors depend on how rich they are in chemicals that can absorb on the metal surface. The earlier publications uncover various parts of the plants, like leaves, root, stem, seed bark, organic product,, mash and so o,n which are significant corrosion inhibitors. Studies of Liu *et al*. [76], report corrosion inhibition qualities of Coptischinensis (CCR) root extricate on aluminium compounds in 3.5% NaCl electrolyte utilizing electrochemical and surface checking strategies which indicated that *C. chinensis* root form a separate interface and mixed type corrosion inhibitor with maximum corrosion inhibition of 94% at 1000 ppm concentration. The corrosion inhibition impact of *Myrmecodia pendans* on carbon steel in 3.5% NaCl indicates corrosion resistivity with a maximum corrosion inhibition of 79.70% at 400 ppm concentration [77]. The impact of Amaranthuscordatus extract on ordinary gentle steel in 0.5 M H_2SO_4 and 1.0 M NaCl has been reported by [78]. The Jatrophacurcas extract was also studied as a corrosion inhibitor for carbon steel in 0.5M H_2SO_4 and 0.5M NaCl using weigh loss techniques. The study demonstrated that *J. Curcas* showed ideal corrosion inhibition of 92.1% (at 1.5 g/L) and 55.5% (2.5 g/L) in 0.5 M NaCl and 0.5 M H_2SO_4 [79]. Recently Kliskic *et al*. [80] studied inhibition properties of aqueous leaf extract of Rosmarinusofficinalis on Al-Mg alloy in 3% NaCl solution utilizing weight reduction, HPLC, EIS and PDP strategies. Apart from the above plant extracts, *Laurus nobilis* L [81], *Citrullus colocynthis* [82], carica papaya [83], Nicotiana Leaves Extract [84], myrrh extract [85], Ginger Extract [86], *Origanum majorana* [87], Aloe Vera [88] and *Vernonia amygdalina* leaf extract [89] and other few have been studied as green corrosion inhibitors for metals and alloys in NaCl solution.

10. ORGANIC COMPOUNDS AS INHIBITORS FOR NACL MEDIA

10.1. Organic Compounds as Inhibitors for Steel, Aluminum and Copper

Utilization of organic compounds is another important technique against metal corrosion in various electrolytic media, including NaCl or ocean water. Organic compounds undergo adsorption on surface of the metal by forming a defensive coating and protect the metal surface against corrosion in marine environments. The functional groups of aliphatic or aromatic organic compounds [90]. Several organic corrosion inhibitors have been reported for steel [91 - 109].

Aluminium exhibits good corrosion resistant due to its property to form a protective oxide layer when exposed to corrosive environments, and this layer of oxide gets soluble in corrosive environments like sea water to accelerate corrosion. Several corrosion inhibitors are developed to protect aluminium and organic corrosion inhibitors have also been explored for their corrosion inhibition properties in sea water [110 - 116].

Due to their superior corrosion resistivity, copper and its alloys have been extensively used in marine environments. Despite having high corrosion resistance, copper, similar to other alloys and metals is corrosion sensitive in ocean waters with chlorides. Organic compounds, particularly heterocyclic compounds are found to be effective as the front line of defence against corrosion of copper in NaCl solution. It is understood that heterocyclic compounds form coordination complexes with Cu(0), Cu(I) and Cu(II) ions which mostly act as a protective barrier for severely corrosive environments having chloride ions. The working of these organic compounds is similar to the mechanism mentioned in the previous part, *i.e.* they get adsorbed on the metal surfaces to prevent them from corrosion [117, 118]. In one of the studies, organic heterocyclic compounds were studied for their corrosion inhibition properties on copper at the concentrations of 0.1 and 0.5 mM in 0.1M NaCl solution [117] and revealed that all compounds exhibited good corrosion inhibition efficiency of above 99% under all study conditions. In another study, 2-amino-5-(ethylthio)-1,3,4-thiadiazole (ATD) [119] was employed as a copper corrosion inhibitor in oxygenated, aerated and deaerated 3% NaCl solution and found that ATD shows a corrosion inhibition efficiency of 94% at 5 mM concentration. The order of corrosion inhibition of ATD is: oxygenated > aerated > de-aerated 3% NaCl solution. Cysteine was used as an environmentally benign corrosion inhibitor for copper [120]. Similarly, benzotriazole (BTAH) and 1-hydroxybenzotriazole (BTAOH) were also investigated on copper in 3% NaCl solution by Finsgar *et al.* [121] using electrochemical methods. Later, sodium succinate (SS) [122] showed an optimum inhibition efficiency of 79.65% at 0.01M concentration for copper. Curkovic *et al.* [123] studied the inhibition effect of three imidazole based organic compounds in 3% NaCl solution.

10.2. Painting

Paint is a mechanical dispersion mixture of one or more pigments in an organic medium called vehicle which is a liquid consisting of non-volatile film-forming material, drying oil and a highly volatile solvent, a thinner. When paint is applied to a metal surface, the thinner evaporates, while the drying oil slowly oxidizes, forming a dry opaque pigmented film. Paint provides protection for metals and alloys from corrosion. Following are some of the fundamental requirements to achieve satisfactory protective coating.

○ Surface preparation.

○ Surface pre-treatment.

○ Anticorrosive or barrier coating application.

○ Antifouling coating application.

CONCLUSION

An insight into marine corrosion, its mechanism, driving forces and preventive measures have been discussed. The main factors in the prevention of marine corrosion are the selection of materials, design, use and maintenance. It has been well understood through a recent survey that, corrosion is found to be responsible for 30% of failures on ships and other marine equipment. There are many methods of preventing marine corrosion like painting, coating, using corrosion inhibitors *etc*. This chapter particularly emphasized on the use of various corrosion inhibitors for preventing marine corrosion. Both organic and inorganic inhibitors are proved to be good corrosion inhibitors with some of the organic inhibitors exhibiting corrosion inhibition efficiency of more 99%. Apart from that, the various types of marine corrosion and the causes have been discussed.

CONSENT FOR PUBLICATION

Not applicable.

CONFLICT OF INTEREST

The author declares no conflict of interest, financial or otherwise.

ACKNOWLEDGEMENTS

Declared none.

REFERENCES

[1]   NACE corrosion cost study, 2002.

[2]   Gupta, R.K.; Malviya, M.; Verma, C.; Quraishi, M.A. Aminoazobenzene and diaminoazobenzene functionalized graphene oxides as a novel class of corrosion inhibitors for mild steel: experimental and DFT studies. *Mater. Chem. Phys.,* **2017**, *198*, 360-373.
       [http://dx.doi.org/10.1016/j.matchemphys.2017.06.030]

[3]   Gupta, R.K.; Malviya, M.; Verma, C.; Gupta, N.K.; Quraishi, M.A. Pyridine-based functionalized graphene oxides as a new class of corrosion inhibitors for mild steel: an experimental and DFT approach. *RSC Advances,* **2017**, *7*, 39063-39074.
       [http://dx.doi.org/10.1039/C7RA05825J]

[4]   Front, Up. The Cost of corrosion in the EEC. *Mater. Perform.,* **1993**, *31*, 3.

[5]   Trethewey, K.R.; Roberge, P.R. Expert overview Corrosion management in the twenty-first century. *Br. Corros. J.,* **1995**, *30*, 192-197.
       [http://dx.doi.org/10.1179/bcj.1995.30.3.192]

[6]   http://www.g2mtlabs.com/corrosion/cost-of-corrosion/

[7]   Roberge, P.R. *Handbook of Corrosion Engineering*; McGraw-Hill, **1999**.

[8]   Yang, Y.; Khan, F.; Thodi, P.; Abbass, R. Corrosion induced failure analysis of subsea pipelines. *Reliab. Eng. Syst. Saf.,* **2017**, *159*, 214-222.
       [http://dx.doi.org/10.1016/j.ress.2016.11.014]

[9]   Verma, C.; Eno, E. Ebenso, Quraishi M A, Ionic liquids as green and sustainable corrosion inhibitors for metals and alloys: an overview. *J. Mol. Liq.,* **2017**, *223*, 403-414.
       [http://dx.doi.org/10.1016/j.molliq.2017.02.111]

[10]   Lua, Y.; Jinga, H.; Hana, Y.; Feng, Z.; Xu, L. Corrosion inhibitors for ferrous and non-ferrous metals and alloys in ionic sodium chloride solutions: A review. *Appl. Surf. Sci.,* **2016**, *389*, 609-622.

[11]   Popov, B.N. Corros. Eng., 2015, 29-92 (b) Ngobiri N C, Oguzie E. E, Oforka N. C, Akaranta O, Comparative study on the inhibitive effect of Sulfadoxine–Pyrimethamine and an industrial inhibitor on the corrosion of pipeline steel in petroleum pipeline water. *Arab. J. Chem.,* **2019**, *12*(7), 1024-1034.

[12]   Southwell, C.R.; Bultman, J.D.; Hummer, C.W. Estimating service life of steel in seawater.*Seawater Corrosion Handbook*; Schumacher, M., Ed.; Noyes Data Corp.: Park Ridge, **1979**, pp. 374-387.

[13]   Centre of Documentation. http://www.cedre.fr/en/spill/erika/erika.php**2013**.

[14]   Canfield, C. *'Super-rust' in Castor leads to calls for more stringent inspections,* **2007**.
       http://www.professionalmariner.com/March-2007/Super-rust-in-Castor-leads-to-calls-for-more-strin
       gent-inspections/(August 2013)

[15]   Howarth, D.J.; Zhang, J. Corrosion: its impact on modern shipping with particular reference to crude oil tankers, **2011**.

[16]   GuedesSoares C. Garbatov Y, Zayed A, Wang G, Corrosion wastage model for ship Crude oil tanks. *Corros. Sci.,* **2008**, *50*, 3095-3106.
       [http://dx.doi.org/10.1016/j.corsci.2008.08.035]

[17]   Melchers, R.E. *Principles of Marine corrosion, Handbook of Ocean Engineering*; Dhanak, M.R.; Xiros, N.I., Eds.; Springer, **2016**.

[18]   Priya, R.; Thyagarajan, K.; Thinaharan, C.; Vijayalakshmi, S.; Ningshen, S. Role of noble metalfission

products on the microstruturalevolutionand corrosion behaviour of ferritic steel-Zr based metal waste formalloys in the simulated repository environment. *J. Alloys Compd.,* **2020**, *820*, 153168. [http://dx.doi.org/10.1016/j.jallcom.2019.153168]

[19]   Li, W.; Nobe, K.; Pearlstein, A.J. Potential/current oscillations and anodic film characteristics of iron in concentrated chloride solutions. *Corros. Sci.,* **1990**, *31*, 615-620. [http://dx.doi.org/10.1016/0010-938X(90)90170-A]

[20]   Zhu, D.; Ooij, W.J. Corrosion protection of AA 2024-T3 by bis-[3 (triethoxysilyl) propyl] tetrasulfide in neutral sodium chloride solution. Part 1: corrosion of AA 2024-T3. *Corros. Sci.,* **2003**, *45*, 2163-2175. [http://dx.doi.org/10.1016/S0010-938X(03)00060-X]

[21]   Moon, S.M.; Pyun, S.I. The corrosion of pure aluminium during cathodic polarization in aqueous solutions. *Corros. Sci.,* **1997**, *39*, 399-408. [http://dx.doi.org/10.1016/S0010-938X(97)83354-9]

[22]   El-Sayed, M. Sherif, Shamy A.M. El, Ramla M. M., Nazhawy A. O.H. El, 5-(Phenyl)-4H-1, 2, 4-triazole-3-thiol as a corrosion inhibitor for copper in 3.5% NaCl solutions. *Mater. Chem. Phys.,* **2007**, *102*, 231-239. [http://dx.doi.org/10.1016/j.matchemphys.2006.12.009]

[23]   Hoepner, T.; Lattemann, S. Chemical impacts from seawater desalination plants-a case study of the northern Red Sea. *Desalination,* **2003**, *152*, 133-140. [http://dx.doi.org/10.1016/S0011-9164(02)01056-1]

[24]   Yan, C.W.; Lin, H.C.; Cao, C.N. Investigation of inhibition of 2-mercaptobenzoxazole for copper corrosion. *Electrochim. Acta,* **2000**, *45*, 2815-2821. [http://dx.doi.org/10.1016/S0013-4686(00)00385-6]

[25]   Schreir, L.L. *Corrosion,* 2nd ed; Newnes- Butterworths: London, **1976**, 1, .

[26]   Laque, F.L. *Marine Corrosion*; Wiley: New York, **1975**.

[27]   Katz, W. *Corrosion Data Sheets (sea water)*; Dechema: Frankfurt am Main, **1976**.

[28]   Schumacher, M. *Sea Water Corrosion Handbook*; Noyes Data Corp.: Park Ridge, NJ, **1979**.

[29]   Bacarella, A.; Land, J.C. GriessJr, The anodic dissolution of copper in flowing sodium chloride solutions between 25 and 175 C. *J. Electrochem. Soc.,* **1973**, *120*, 459. [http://dx.doi.org/10.1149/1.2403477]

[30]   Kester, D.R. Chemistry of iron in marine systems. *Nav. Res. Rev.,* **1974**, (9), 3.

[31]   Atkinson, G.; Gilligan, T.J. *Nav. Res. Rev.,* **1974**, (9), 17.

[32]   Foley, R.T.; Gilligan, T.J. *'Complex ions and corrosion', Report AD AOll099*; NTIS: Springfield, VA, **1975**.

[33]   DEXTER, S.J.; Culberson, C. Global variability of natural sea water[J]. *Mater. Perform.,* **1980**, *19*(9), 16.

[34]   Compton K.G., Proc. Int. Corrosion Forum, Toronto, 1975, NACE, paper 13..

[35]   Castle, J.E.; Pepler, D.C.; Peplov, D.B. ESCA investigation of iron-rich protective films on aluminium brass condenser tubes. *Corros. Sci.,* **1976**, *16*, 145. [http://dx.doi.org/10.1016/0010-938X(76)90055-X]

[36]   Castle, J.E.; Chamberlain, A.H.L; Garner, B.; Sadegh, P.M.; Aladjen, A. *The use of synthetic environments for corrosion testing,* **1988**, , 174.

[37]   Chandler, H.E. Met. Prog., June 1979, 47.

[38]   Houghton, D.R.; Gage, S.A. Trans. Insf. Mar. Eng. (TM),1979, 91, (11), paper 13.

[39]   R. Mitchell. *Nav. Res. Rev.,* **1972**, *25*, 11.

[40]   R. c. SALVAREZZA.M. F. L. de MELE,and H. A. *VIDELA:Corrosion,* **1983**, *39*, 26.

[41]   G. R. Weber. *Mater. Perform.,* **1983**, *23*(10), 24.

[42]   Scotto, V.; Dicinto, R.; Marcenaro, G. The influence of marine aerobic microbial film on stainless steel corrosion behaviour. *Corros. Sci.,* **1985**, *25*, 185-194.
[http://dx.doi.org/10.1016/0010-938X(85)90094-0]

[43]   Scotto, V.; Alabiso, G.; Marcenaro, G. An example of microbiologically influenced corrosion: The behaviour of stainless steels in natural seawater. *Bioelectrochem. Bioenerg.,* **1986**, *16*, 347-355.
[http://dx.doi.org/10.1016/0302-4598(86)85014-0]

[44]   Johnsen, R.; Bardal, E.; Marcenaro, G. Proc. Int. Corrosion Forum, 1986, Houston, TX, NACE, paper 227..

[45]   Ijsseling, F.P.; Drolenga, L.J.P.; Kolster, B.H. Influence of temperature on corrosion product film formation on CuNi10Fe in the low temperature range: i. corrosion rate as a function of temperature in well aerated sea water. *Br. Corros. J.,* **1982**, *17*, 162-167.
[http://dx.doi.org/10.1179/000705982798274282]

[46]   J. M. Krougman and F. P. Ijsseling. *Proc. 4th Int. Congo on 'Marine corrosion and fouling', Juan-le--Prins*; Comite International Permanent pour la Recherchesur la Preservation des Materiaux au Milieu Marin: Paris, **1976**, p. 297.

[47]   Parker, J.G.; Roscow, J.A. Proc. 8th Int. *Congo on Metallic Corrosion, Mainz,* **1981**, *1285*, 1981.

[48]   Efird, K.D.; Lee, T.S. Putrid sea water as a corrosive medium. *Corrosion,* **1979**, *35*, 79-83.
[http://dx.doi.org/10.5006/0010-9312-35.2.79]

[49]   Smith, A.L. *Reliability of engineering materials*; Butterworths: London, **1984**, pp. 123-147.
[http://dx.doi.org/10.1016/B978-0-408-01507-3.50011-6]

[50]   Kato, C.; Pickering, H.W.; Castle, J.E. Effect of sulfide on the corrosion of cu9.4 ni1.7 fe alloy in aqueous NaCl solution. *J. Electrochem. Soc.,* **1984**, *131*, 1226.
[http://dx.doi.org/10.1149/1.2115792]

[51]   Roberge, P.R. *Corrosion Engineering: Principles and Practice*; McGraw-Hill: New York, **2008**.

[52]   Jotun: Training Course Handouts. Jotun Marine Coatings, 2001.

[53]   Standard, Norsok Cathodic Protection (M-503). Edition 3, May 2007.

[54]   Taylan, M. GEM 418E Corrosion and Corrosion in Marine Environment. Istanbul Technical University, Department of Naval Architecture and Marine Engineering, GEM 418E Lecture Notes, Istanbul 2009.

[55]   Ashworth, V. Principles of Cathodic Protection. http://www.elsevierdirect.com/brochures/shreir /PDF/Principles_of_Cathodic_Protection.pdf

[56]   Bushman, J.B. Corrosion and cathodic protection theory. http://www.bushman.cc/pdf/corrosion_ theory.pdf

[57]   a) Aballe, A.; Bethencourt, M.; Botana, F.J.; Marcos, M. CeCl$_3$ and LaCl$_3$ binary solutions as environment-friendly corrosion inhibitors of AA5083 Al–Mg alloy in NaCl solutions. *J. Alloys Compd.,* **2001**, *323–324*, 855-858.
[http://dx.doi.org/10.1016/S0925-8388(01)01160-4] b) Chaudhry, A.U.; Mittal, V.; Mishra, B. Nano nickel ferrite (NiFe$_2$O$_4$) as anti-corrosion pigment for API 5L X-80 steel: An electrochemical study in acidic and saline media. *Dyes Pigm.,* **2015**, *118*, 18-26.
[http://dx.doi.org/10.1016/j.dyepig.2015.02.023]

[58]   Mishra, A.K.; Balasubramaniam, R. Corrosion inhibition of aluminium by rare earth chlorides. *Mater. Chem. Phys.,* **2007**, *103*, 385-393.
[http://dx.doi.org/10.1016/j.matchemphys.2007.02.079]

[59]   Aramaki, K. Self-healing mechanism of an organosiloxane polymer film containing sodium silicate and cerium (III) nitrate for corrosion of scratched zinc surface in 0.5 M NaCl. *Corros. Sci.,* **2002**, *44*, 1621-1632.
[http://dx.doi.org/10.1016/S0010-938X(01)00171-8]

[60]   Aramaki, K. Preparation of chromate-free, self-healing polymer films containing sodium silicate on zinc pretreated in a cerium (III) nitrate solution for preventing zinc corrosion at scratches in 0.5M NaCl. *Corros. Sci.,* **2002**, *44*, 1375-1389.
[http://dx.doi.org/10.1016/S0010-938X(01)00138-X]

[61]   Naderi, R.; Mahdavian, M.; Attar, M.M. Electrochemical behavior of organic and inorganic complexes of Zn (II) as corrosion inhibitors for mild steel: Solution phase study. *Electrochim. Acta,* **2009**, *54*, 6892-6895.
[http://dx.doi.org/10.1016/j.electacta.2009.06.073]

[62]   Fouda, A.S.; Megahed, H.; Ead, D.M. Lanthanides as environmentally friendly corrosion inhibitors of iron in 3.5% NaCl solution. *Desalination Water Treat.,* **2013**, *51*, 3164-3178.
[http://dx.doi.org/10.1080/19443994.2012.749023]

[63]   Aramaki, K. Synergistic inhibition of zinc corrosion in 0.5 M NaCl by combination of cerium (III) chloride and sodium silicate. *Corros. Sci.,* **2002**, *44*, 871-886.
[http://dx.doi.org/10.1016/S0010-938X(01)00087-7]

[64]   Aramaki, K. The inhibition effects of cation inhibitors on corrosion of zinc in aerated 0.5 M NaCl. *Corros. Sci.,* **2001**, *43*, 1573-1588.
[http://dx.doi.org/10.1016/S0010-938X(00)00144-X]

[65]   Mishra, A.K.; Balasubramaniam, R. Corrosion inhibition of aluminum alloy AA 2014 by rare earth chlorides. *Corros. Sci.,* **2007**, *49*, 1027-1044.
[http://dx.doi.org/10.1016/j.corsci.2006.06.026]

[66]   Arenas, M.A.; Bethencourt, M.; Botana, F.J.; de Damborenea, J.; Marcos, M. Inhibition of 5083 aluminium alloy and galvanised steel by lanthanide salts. *Corros. Sci.,* **2001**, *43*, 157-170.
[http://dx.doi.org/10.1016/S0010-938X(00)00051-2]

[67]   Bethencourt, M.; Botana, F.J.; Cauqui, M.A.; Marcos, M.; Rodrıguez, M.A.; Rodríguez-Izequierdo, J.M. Protection against corrosion in marine environments of AA5083 Al–Mg alloy by lanthanide chlorides. *J. Alloys Compd.,* **1997**, *250*, 455-460.
[http://dx.doi.org/10.1016/S0925-8388(96)02826-5]

[68]   Matter, E.A.; Kozhukharov, S.; Machkova, M.; Kozhukharov, V. Electrochemical studies on the corrosion inhibition of AA2024 aluminium alloy by rare earth ammonium nitrates in 3.5% NaCl solutions. *Mater. Corros.,* **2013**, *64*, 408-414.
[http://dx.doi.org/10.1002/maco.201106349]

[69]   Eldwaib, K.A. *Env. Biol. Sci.,* **2013**, *1*, 2320-4079.

[70]   Davó, B.; de Damborenea, J.J. Use of rare earth salts as electrochemical corrosion inhibitors for an Al–Li–Cu (8090) alloy in 3.56% NaCl. *Electrochim. Acta,* **2004**, *49*, 4957-4965.
[http://dx.doi.org/10.1016/j.electacta.2004.06.008]

[71]   Davó, B.; Conde, A.; de Damborenea, J.J. Inhibition of stress corrosion cracking of alloy AA8090 T-8171 by addition of rare earth salts. *Corros. Sci.,* **2005**, *47*, 1227-1237.
[http://dx.doi.org/10.1016/j.corsci.2004.07.028]

[72]   Aramaki, K. Treatment of zinc surface with cerium (III) nitrate to prevent zinc corrosion in aerated 0.5 M NaCl. *Corros. Sci.,* **2001**, *43*, 2201-2215.
[http://dx.doi.org/10.1016/S0010-938X(00)00189-X]

[73]   Aramaki, K. The inhibition effects of chromate-free, anion inhibitors on corrosion of zinc in aerated 0.5 M NaCl. *Corros. Sci.,* **2001**, *43*, 591-604.
[http://dx.doi.org/10.1016/S0010-938X(00)00085-8]

[74] Aramaki, K. Cerium (III) chloride and sodium octylthiopropionate as an effective inhibitor mixture for zinc corrosion in 0.5 M NaCl. *Corros. Sci.,* **2002**, *44*, 1361-1374.
[http://dx.doi.org/10.1016/S0010-938X(01)00116-0]

[75] W. A. Badawy W. A. Ismail K. M, Fathi A. M, Effect of Ni content on the corrosion behavior of Cu–Ni alloys in neutral chloride solutions. *Electrochim. Acta,* **2005**, *50*, 3603-3608.
[http://dx.doi.org/10.1016/j.electacta.2004.12.030]

[76] Liu, W.; Singh, A.; Lin, Y.; Ebenso, E.E.; Tianhan, G.; Ren, C. Corrosion inhibition of Al-alloy in 3.5% NaCl solution by a natural inhibitor: an electrochemical and surface study. *Int. J. Electrochem. Sci.,* **2014**, *9*, 5560-5573.

[77] Pradityana, A; Shahab, S A; Noerochim, L; Susanti, D Inhibition of corrosion of carbon steel in 3.5% NaCl solution by MyrmecodiaPendans extract *Inter. J. Corros,* **2016**, 1-6.

[78] Nwankwo, M.O.; Offor, P.O.; Neife, S.I.; Oshionwu, L.C.; Idenyi, N.E. Amaranthuscordatus as a Green Corrosion Inhibitor for Mild Steel in H_2SO_4 and NaCl. *J. Miner. Mater. Charact. Eng.,* **2014**, *2*, 194-199.
[http://dx.doi.org/10.4236/jmmce.2014.23024]

[79] Omotoyinbo, J.A.; Oloruntoba, D.T.; Olusegun, S.J. Corrosion inhibition of pulverized jatrophacurcas leaves on medium carbon steel in 0.5 M H_2SO_4 and NaCl environments. *Inter. J. Sci. Tech.,* **2013**, *2*, 510-514.

[80] Kliskic, M.; Radosevic, J.; Gudic, S.; Katalinic, V. Aqueous extract of Rosmarinusofficinalis L. as inhibitor of Al–Mg alloy corrosion in chloride solution. *J. Appl. Electrochem.,* **2000**, *30*, 823-830.
[http://dx.doi.org/10.1023/A:1004041530105]

[81] Halambek, J.; Berkovic, K.; Vorkapic-Furac, J.; Laurusnobilis, L. oil as green corrosion inhibitor for aluminium and AA5754 aluminium alloy in 3% NaCl solution. *Mater. Chem. Phys.,* **2013**, *137*, 788-795.
[http://dx.doi.org/10.1016/j.matchemphys.2012.09.066]

[82] Al-Dokheily, M.E.; Kredy, H.M.; Al-Jabery, R.N. Inhibition of copper corrosion in H_2SO_4, NaCl and NaOH solutions by Citrulluscolocynthis fruits extract. *J. Natur. Sci. Res.,* **2014**, *4*, 60-73.

[83] Loto, C.A.; Popoola, A.P.I. Electrochemical potential monitoring of corrosion and inhibitors protection of mild steel embedded in concrete in NaCl solution. *J. Mater. Environ. Sci.,* **2012**, *3*, 816-825.

[84] Abd-El-Khalek, D.E.; Abd-El-Nabey, B.A.; Abdel-Gaber, A.M. Evaluation of nicotiana leaves extract as corrosion inhibitor for steel in acidic and neutral chloride solutions, PortugaliaeElectrochim. *Acta,* **2012**, *30*(4), 247-259.

[85] Gadow, H.S.; Motawea, M.M.; Elabbasy, H.M. Investigation of myrrh extract as a new corrosion inhibitor for α-brass in 3.5% NaCl solution polluted by 16 ppm sulfide. *RSC Advances,* **2017**, *7*, 29883-29898.
[http://dx.doi.org/10.1039/C7RA04271J]

[86] A. El-Aziz S. Fouda. Abdel Nazeer A, Ibrahim M, Fakih M, Ginger extract as green corrosion inhibitor for steel in sulfide polluted salt water. *J. Korean Chem. Soc.,* **2013**, *57*, 272-278.
[http://dx.doi.org/10.5012/jkcs.2013.57.2.272]

[87] Challouf, H.; Souissi, N.; Ben Messaouda, M.; Abidi, R.; Madani, A. Origanummajorana Extracts as Mild Steel Corrosion Green Inhibitors in Aqueous Chloride Medium. *J. Environ. Prot. (Irvine Calif.),* **2016**, *7*, 532-544.
[http://dx.doi.org/10.4236/jep.2016.74048]

[88] Al–Asadi, A.A.; Abdullah, A.S.; Khaled, N.I.; Alkhafaja, R.J.M. Corrosion inhibitors for ferrous and non-ferrous metals and alloys in ionic sodium chloride solutions: A review. *Inter. Res. J. Eng. Tech.,* **2015**, *2*, 604-607.

[89]   Ahmed, A.; Siyaka, S.O. A Study of VeroniaAmygdalina Leaf Extract on Corrosion Resistance of 96% Al, 3.5% Zn and 0.5% Mg in Nacl Solution. *Bioprocess Eng.,* **2017**, *1*, 77-80.

[90]   a) Prabakaran, M.; Kim, S.; Mugila, N.; Hemapriya, V.; Parameswari, K.; Chitra, S.; Chung, I.M. Aster koraiensis as nontoxic corrosion inhibitor for mild steel in sulfuric acid. *J. Ind. Eng. Chem.,* **2017**, *52*, 235-242
[http://dx.doi.org/10.1016/j.jiec.2017.03.052] b) Rouzmeh, S.S.; Naderi, R.; Mahdavian, M. A sulfuric acid surface treatment of mild steel for enhancing the protective properties of an organosilane coating. *Prog. Org. Coat.,* **2017**, *103*, 156-164.
[http://dx.doi.org/10.1016/j.porgcoat.2016.10.033] c) Verma, C.; Quraishi, M.A.; Olasunkanmi, L.O.; Eno, E. Ebenso. L-Proline-promoted synthesis of 2-amino-4-arylquinoline-3-carbonitriles as sustainable corrosion inhibitors for mild steel in 1 M HCl: experimental and computational studies. *RSC Advances,* **2015**, *5*, 85417-85430.
[http://dx.doi.org/10.1039/C5RA16982H]

[91]   Amar, H.; Benzakour, J.; Derja, A.; Villemin, D.; Moreau, B.; Braisaz, T. Piperidin-1-yl-phosphonic acid and (4-phosphono-piperazin-1-yl) phosphonic acid: a new class of iron corrosion inhibitors in sodium chloride 3% media. *Appl. Surf. Sci.,* **2006**, *252*, 6162-6172.
[http://dx.doi.org/10.1016/j.apsusc.2005.07.073]

[92]   Amar, H.; Benzakour, J.; Derja, A.; Villemin, D.; Moreau, B. A corrosion inhibition study of iron by phosphonic acids in sodium chloride solution. *J. Electroanal. Chem. (Lausanne Switz.),* **2003**, *558*, 131-139.
[http://dx.doi.org/10.1016/S0022-0728(03)00388-7]

[93]   Sahin, M.; Bilgic, S.; Yilmaz, H. The inhibition effects of some cyclic nitrogen compounds on the corrosion of the steel in NaCl mediums. *Appl. Surf. Sci.,* **2002**, *195*, 1-7.
[http://dx.doi.org/10.1016/S0169-4332(01)00783-8]

[94]   Singh, A.; Lin, Y ; Quraishi, M.A.; Olasunkanmi, L.O.; Fayemi, O.E.; Sasikumar, Y.; Ramaganthan, B.; Bahadur, I.; Obot, I.B.; Adekunle, A.S.; Kabanda, M.M.; Ebenso, E.E. Porphyrins as corrosion inhibitors for N80 Steel in 3.5% NaCl solution: Electrochemical, quantum chemical, QSAR and Monte Carlo simulations studies. *Molecules,* **2015**, *20*(8), 15122-15146.
[http://dx.doi.org/10.3390/molecules200815122] [PMID: 26295223]

[95]   Solehudin, A. Performance of benzotriazole as corrosion inhibitors of carbon steel in chloride solution containing hydrogen sulfide. *Inter. Refer. J. Eng. Sci.,* **2012**, *1*, 21-26.

[96]   Zhang, X.; Wang. F.; He, Y.; Du, Y. Study of the inhibition mechanism of imidazoline amide on CO2 corrosion of Armco iron. *Corros. Sci.,* **2001**, *43*, 1417-1431.
[http://dx.doi.org/10.1016/S0010-938X(00)00160-8]

[97]   Amar, H.; Braisaz, T.; Villemin, D.; Moreau, B. Thiomorpholin-4-ylmethyl-phosphonic acid and morpholin-4-methyl-phosphonic acid as corrosion inhibitors for carbon steel in natural seawater. *Mater. Chem. Phys.,* **2008**, *110*, 1-6.
[http://dx.doi.org/10.1016/j.matchemphys.2007.10.001]

[98]   Jiang, X.; Zheng, Y.G.; Ke, W. Effect of flow velocity and entrained sand on inhibition performances of two inhibitors for CO_2 corrosion of N80 steel in 3% NaCl solution. *Corros. Sci.,* **2005**, *47*, 2636-2658.
[http://dx.doi.org/10.1016/j.corsci.2004.11.012]

[99]   Amar, H.; Benzakour, J.; Derja, A.; Villemin, D.; Moreau, B.; Braisaz, T.; Tounsi, A. Synergistic corrosion inhibition study of Armco iron in sodium chloride by piperidin-1-yl-phosphonic acid–Zn^{2+} system. *Corros. Sci.,* **2008**, *50*, 124-130.
[http://dx.doi.org/10.1016/j.corsci.2007.06.010]

[100]   Okafor, P.C.; Liu, X.; Zheng, Y.G. Corrosion inhibition of mild steel by ethylaminoimidazoline derivative in CO_2-saturated solution. *Corros. Sci.,* **2009**, *51*, 761-768.
[http://dx.doi.org/10.1016/j.corsci.2009.01.017]

[101]   Gece, G.; Bilgiç, S. Quantum chemical study of some cyclic nitrogen compounds as corrosion inhibitors of steel in NaCl media. *Corros. Sci.,* **2009**, *51*, 1876-1878.
[http://dx.doi.org/10.1016/j.corsci.2009.04.003]

[102]   El-Sayed, M. Sherif, Erasmus R. M, Comins J. D, *In situ* Raman spectroscopy and electrochemical techniques for studying corrosion and corrosion inhibition of iron in sodium chloride solutions. *Electrochim. Acta,* **2010**, *55*, 3657-3663.
[http://dx.doi.org/10.1016/j.electacta.2010.01.117]

[103]   Lopez, D.A.; Simison, S.N.; de Sanchez, S.R. The influence of steel microstructure on CO_2 corrosion. EIS studies on the inhibition efficiency of benzimidazole. *Electrochim. Acta,* **2003**, *48*, 845-854.
[http://dx.doi.org/10.1016/S0013-4686(02)00776-4]

[104]   Sekine, I.; Hirakawa, Y. Effect of 1-hydroxyethylidene-1, 1-diphosphonic acid on the corrosion of SS 41 steel in 0.3% sodium chloride solution. *Corrosion,* **1986**, *42*, 272-277.
[http://dx.doi.org/10.5006/1.3584904]

[105]   Chong, A.L.; Mardel, J.I.; MacFarlane, D.R.; Forsyth, M. A.y E. Somers, Synergistic corrosion inhibition of mild steel in aqueous chloride solutions by an imidazolinium carboxylate salt. *ACS Sustain. Chem.& Eng.,* **2016**, *4*, 1746-1755.
[http://dx.doi.org/10.1021/acssuschemeng.5b01725]

[106]   Bokati, K.S.; Dehghanian, C.; Yari, S. Corrosion inhibition of copper, mild steel and galvanically coupled copper-mild steel in artificial sea water in presence of 1H-benzotriazole, sodium molybdate and sodium phosphate. *Corros. Sci.,* **2017**, *126*, 272-285.
[http://dx.doi.org/10.1016/j.corsci.2017.07.009]

[107]   Alinejad, S.; Naderi, R.; Mahdavian, M. Effect of inhibition synergism of zinc chloride and 2-mercaptobenzoxzole on protective performance of an ecofriendly silane coating on mild steel. *J. Ind. Eng. Chem.,* **2017**, *48*, 88-98.
[http://dx.doi.org/10.1016/j.jiec.2016.12.024]

[108]   Aloysius, A.; Ramanathan, R.; Christy, A.; Baskaran, S.; Antony, N. Experimental and theoretical studies on the corrosion inhibition of vitamins–Thiamine hydrochloride or biotin in corrosion of mild steel in aqueous chloride environment. *Egyptian J. Petro,* **2018**, *27*, 371-381.
[http://dx.doi.org/10.1016/j.ejpe.2017.06.003]

[109]   a) Xhanari, K.; Finsgar, M. Organic corrosion inhibitors for aluminum and its alloys in chloride and alkaline solutions: a review. *Arab. J. Chem.,* **2019**, *12*, 4646-4663.
[http://dx.doi.org/10.1016/j.arabjc.2016.08.009] b) Zaid, B.; Maddache, N.; Saidi, D.; Souami, N.; Bacha, N.; Si Ahmed, A. Electrochemical evaluation of sodium metabisulfite as environmentally friendly inhibitor for corrosion of aluminum alloy 6061 in a chloride solution. *J. Alloys Compd.,* **2015**, *629*, 188-196.
[http://dx.doi.org/10.1016/j.jallcom.2015.01.003]

[110]   a) Zhu, D.; Ooij, W.J. Corrosion protection of AA 2024-T3 by bis-[3-(triethoxysilyl) propyl] tetrasulfide in sodium chloride solution.: Part 2: mechanism for corrosion protection. *Corros. Sci.,* **2003**, *45*, 2177-2197.
[http://dx.doi.org/10.1016/S0010-938X(03)00061-1] b) Migahed, M.A.; El-Rabiei, M.M.; Nady, H.; Fathy, M. The synthesis and characterization of benzotriazole-based cationic surfactants and the evaluation of their corrosion inhibition efficiency on copper in seawater. *J. Environ. Chem. Eng.,* **2016**, *4*, 3741-3752.
[http://dx.doi.org/10.1016/j.jece.2016.08.020]

[111]   Osman, M.M. Corrosion inhibition of aluminium–brass in 3.5% NaCl solution and sea water. *Mater. Chem. Phys.,* **2001**, *71*, 12-16.
[http://dx.doi.org/10.1016/S0254-0584(00)00510-1]

[112]   Sherif, E.M. Su-Moon Park, Effects of 1, 4-naphthoquinone on aluminum corrosion in 0.50 M sodium chloride solutions. *Electrochim. Acta,* **2006**, *51*, 1313-1321.

[http://dx.doi.org/10.1016/j.electacta.2005.06.018]

[113] Zheludkevich, M.L.; Yasakau, K.A.; Poznyak, S.K. Ferreira\ M G S, Triazole and thiazole derivatives as corrosion inhibitors for AA2024 aluminium alloy. *Corros. Sci.,* **2005**, *47*, 3368-3383.
[http://dx.doi.org/10.1016/j.corsci.2005.05.040]

[114] Lamaka, S.V.; Zheludkevich, M.L.; Yasakau, K.A.; Montemor, M.F.; Ferreira, M.G.S. High effective organic corrosion inhibitors for 2024 aluminium alloyElectrochim. *Acta,* **2007**, *52*, 7231-7247.

[115] Zhu, D.; Ooij, W.J. Enhanced corrosion resistance of AA 2024-T3 and hot-dip galvanized steel using a mixture of bis-[triethoxysilylpropyl] tetrasulfide and bis-[trimethoxysilylpropyl] amineElectrochim. *Acta,* **2004**, *49*, 1113-1125.
[http://dx.doi.org/10.1016/j.electacta.2003.10.023]

[116] Song-mei, L.; Hong-rui, Z.; Jian-hua, L. Corrosion behavior of aluminum alloy 2024-T3 by 8-hydroxy-quinoline and its derivative in 3.5% chloride solution. *Trans. Nonferrous Met. Soc. China,* **2007**, *17*, 318-325.
[http://dx.doi.org/10.1016/S1003-6326(07)60092-2]

[117] Sherif, E.M. Effects of 2-amino-5-(ethylthio)-1, 3, 4-thiadiazole on copper corrosion as a corrosion inhibitor in 3% NaCl solutions. *Appl. Surf. Sci.,* **2006**, *252*, 8615-8623.
[http://dx.doi.org/10.1016/j.apsusc.2005.11.082]

[118] Amin, M.A. Weight loss, polarization, electrochemical impedance spectroscopy, SEM and EDX studies of the corrosion inhibition of copper in aerated NaCl solutions. *J. Appl. Electrochem.,* **2006**, *36*, 215-226.
[http://dx.doi.org/10.1007/s10800-005-9055-1]

[119] Zucchi, F.; Trabanelli, G.; Monticelli, C. The inhibition of copper corrosion in 0.1 M NaCl under heat exchange conditionsCorros. *Sci.,* **1996**, *38*, 147-154.

[120] Ismail, K.M. Evaluation of cysteine as environmentally friendly corrosion inhibitor for copper in neutral and acidic chloride solutionsElectrochim. *Acta,* **2007**, *52*, 7811-7819.

[121] Finsgar, M.; Lesar, A.; Kokalj, A.; Milosev, I. A comparative electrochemical and quantum chemical calculation study of BTAH and BTAOH as copper corrosion inhibitors in near neutral chloride solutionElectrochim. *Acta,* **2008**, *53*, 8287-8297.
[http://dx.doi.org/10.1016/j.electacta.2008.06.061]

[122] Hazzazi, O.A. Corrosion inhibition studies of copper in highly concentrated NaCl solutions. *J. Appl. Electrochem.,* **2007**, *37*, 933-940.
[http://dx.doi.org/10.1007/s10800-007-9332-2]

[123] Curkovic, H.O.; Stupnisek-Lisac, E.; Takenouti, H. Electrochemical quartz crystal microbalance and electrochemical impedance spectroscopy study of copper corrosion inhibition by imidazoles. *Corros. Sci.,* **2009**, *51*, 2342-2348.
[http://dx.doi.org/10.1016/j.corsci.2009.06.018]

SUBJECT INDEX

N. Suresh Kumar, P. Banerjee, H. Manjunatha and K. Chandra Babu Naidu (Eds.)
All rights reserved-© 2021 Bentham Science Publishers

Z

Zinc 69
    metaphosphate 69
    orthophosphate 69
Zinc metal 21, 20
    silicates 21
    pigmented 21